SCIENCE, SOCIETY, AND THE SUPERMARKET

SCIENCE, SOCIETY, AND THE SUPERMARKET

The Opportunities and Challenges of Nutrigenomics

David Castle, PhD
University of Ottawa
Ottawa, Ontario, Canada

Cheryl Cline, PhD
Abdallah S. Daar, D. Phil
Charoula Tsamis, MA
Peter A. Singer, MD
University of Toronto
Toronto, Ontario, Canada

BICENTENNIAL
1807
WILEY
2007
BICENTENNIAL

WILEY-INTERSCIENCE
A JOHN WILEY & SONS, INC., PUBLICAION

For general information on our other products and services or for technical support, please contact our Customer Care Department within the United States at 877-762-2974, outside the United States at 317-572-3993 or fax 317-572-4002.

Wiley also publishes its books in a variety of electronic formats. Some content that appears in print may not be available in electronic formats. For more information about Wiley products, visit our web site at www.wiley.com.

Library of Congress Cataloging-in-Publication Data:

Science, society, and the supermarket : the opportunities and challenges
   of nutrigenomics / by David Castle. . . [et al.].
        p.   cm.
   Includes bibliographical references and index.
   ISBN 13: 978-0-471-77000-8
   ISBN 10: 0-471-77000-0 (cloth)
      1. Nutrition—Genetic aspects.   2. Physiological genomics.   3. Nutrient interactions.
   I. Castle, David, 1967–

QP144.G45S35 2006
612.3—dc22   2006044044

Printed in the United States of America

10 9 8 7 6 5 4 3 2 1

CONTENTS

**3 THE ETHICS OF NUTRIGENOMIC TESTS AND
INFORMATION   49**

4 ALTERNATIVES FOR NUTRIGENOMIC SERVICE DELIVERY   77

PREFACE

Rarely does a day pass when there is not some news about the latest fad diet, the latest passé diet, research about some link between nutrition and disease, reports of the health-enhancing properties of certain foods, revisions to food guides, reversals of opinion about dietary supplements, and so on. Why so much information, and why so much interest in it? The reason is obvious: eating and drinking are necessary, highly intimate, richly cultural, and potentially pleasurable activities. Despite the mindlessness with which we sometimes hurriedly feed ourselves with banal food during the rush of the day, no one will deny on reflection the profound meaning and implications of our food choices. What we consume can make us healthy, happy, and part of a community. It can also reverse all of these.

It is said that the code of human life has been cracked, by which it is meant that the mysteries of human biology are being unraveled like DNA's double helix. It is also said that nutrition has gone molecular, meaning that nutrition has become an experimental science of life's chemicals. Nutritional genomics attempts to combine human genomics with the nutritional sciences. Nutrigenomics, as it is widely known, attempts to understand how nutrients in our food interact with our genes and how small genetic differences can have important implications for how our bodies use nutrients.

Nutrigenomics is a young science full of potential. As research programs mature, and as the applications of the science are implemented, a number of serious questions are being raised: Can a "gene-based diet" exist? Are consumers being defrauded by bogus services? Should genetic tests be offered over the internet? Is it right to make a recommendation if there's a chance that tomorrow's science will contradict today's advice? Who will get access to the nutrigenomics? In this book we consider many ethical and legal issues that arise in nutrigenomics, and provide recommendations about how these social issues might be addressed as nutrigenomics develops.

We consider how health information and biological samples should be handled, how nutrigenomics can be offered to the public, how the field might be regulated, and we address the need for equitable access to the beneficial applications of nutrigenomics. These, one might say, are bread and butter issues for bioethics, and we agree. This is the first sustained discussion of the ethical and legal issues generated by nutrigenomics, one which identifies, analyzes, and makes recommendations about the core issues. We encourage the reader to think about these issues in light of a broader context about our changing understanding of food. That context draws our attention to the potential to "medicalize" food, put health and pleasure in juxtaposition, assign personal responsibility for dietary habits and their consequences, strengthen the role of genetics in health, and ensure equitable access not only to food but to knowledge about food.

August 2006   DAVID CASTLE

ACKNOWLEDGMENTS

Expert Panel

George Gaskell
Professor, Department of Social Psychology, London School of Economics

Peter Gillies
Cross-Appointed Professor, Department of Nutritional Sciences, University of Toronto

Jean Jones
Chair, Health Council, Consumers' Association of Canada

Bartha Knoppers
Professor, Faculty of Law, Université de Montréal

John Milner
Chief, Nutritional Science Research Group, National Cancer Institute

Barbara Schneeman
Professor, Nutrition Department, University of California, Davis

Patrick Terry
President, International Genetic Alliance

Professional Assistance

Michael Keating
Editor

Jeremy Willard
Summer Student Assistant

Funding

The Canadian Program on Genomics and Global Health is supported primarily by Genome Canada through the Ontario Genomics Institute and the Ontario Research and Development Challenge Fund. Matching partners are listed at www.geneticsethics.net. Peter Singer is supported by a Canadian Institutes of Health Research Distinguished Investigator award. Abdallah Daar is supported by the McLaughlin Centre for Molecular Medicine.

1

NUTRITIONAL GENOMICS: OPPORTUNITIES AND CHALLENGES

Let food be your medicine and medicine be your food.
—Hippocrates, the father of modern medicine, ca. 400 B.C.

1.1 INTRODUCTION

The concept that some foods not only provide sustenance but also have medicinal properties is not a new one. In many cultures, food and medicine historically shared common origins (Verschuren 2002). Although a food-based approach to wellness was eclipsed in a number of countries by what we call modern medicine, with its use of drugs targeted at fighting specific diseases, the role of diet in health promotion and disease prevention is once again becoming popular (Hasler 2000). Health food and organic foods are gaining market share in many countries. About 40 percent of U.S. super-market companies offer programs to help people manage diseases, and about half their stores draw a direct connection between particular conditions and the consumption of certain foods (Sloan 2000). In Canada, many people regularly take vitamins, minerals, herbal products, homeopathic medicines, or similar substances, creating a multibillion dollar a year market. In a recent

Science, Society, and the Supermarket: The Opportunities and Challenges of Nutrigenomics,
By David Castle, Cheryl Cline, Abdallah S. Daar, Charoula Tsamis, and Peter A. Singer
Copyright © 2007 John Wiley & Sons, Inc.

survey of Canadians, 46 percent of women and 33 percent of men reported using at least one natural health product (Rowe and Toner 2003). Of those taking a supplement, 38 percent took vitamin and mineral supplements, and 15 percent used herbal products (Troppman 2002), a statistic corroborated by more recent provincial data (Mendelson et al. 2003).

Even before the genomics revolution of recent years, we knew of such gene–diet interactions as lactose intolerance (inability to digest sugars found in milk), alcohol dehydrogenase deficiency (inability to digest alcohol), phenylketonuria (a condition that renders infants vulnerable to brain damage if they eat certain common foods), and differences in blood lipid profiles and various health outcomes linked to the consumption of high-fat diets (Fogg-Johnson and Merolli 2000). Until recently, scientists lacked the tools to fully understand the underlying mechanisms that cause these conditions. In 2001, the completion of the Human Genome Project, the mapping and sequencing of most of the human genome, opened the door to a detailed examination of how our genes interact with a number of environmental factors, including our foods. As scientists probe deeper into the human genome, we are learning more about how genetic differences play a role in a number of conditions and diseases, such as diabetes, obesity, cancer, heart disease, birth defects, and food allergies. Nutrients may also modulate the expression of our genes and thus influence the way our bodies function. Current scientific thinking holds that these processes contribute with other factors to the initiation, development, or progression of disease. Human genomics has also made it possible to identify molecular targets for dietary components and to better understand physiological mechanisms that can be modulated by diet in order to improve health and well-being (Milner 2002).

The science of nutritional genomics has evolved in the midst of a paradigm shift toward individualized medicine. Individualized medicine is based on the idea that genetic differences between people matter greatly in the prevention, development, and treatment of disease. This idea has propelled the field of *pharmacogenomics*, the study of individual variability in drug response. *Nutritional genomics* is the food analog of pharmacogenomics: It promises to help improve human health by revealing the links between our food and how our bodies work at the molecular level (Simopoulos 2004). The applications of nutritional genomics include the development of gene-based tests for chronic diseases influenced by diet, the refinement of dietary recommendations, the improvement in the risk–benefit assessment of foods, and the substantiation of health claims.

Like any new field of science and technology, nutritional genomics faces a number of challenges. These are of two kinds: scientific and technological challenges; and broadly speaking, social challenges. The scientific and technological challenges for nutritional genomics revolve around the need to provide the scientific evidence of safe and effective health benefits that are not yet readily available. The social challenges pertain to the range of ethical, economic, and legal issues about how best to interpret, utilize, and deliver this new knowledge. These goals involve a number of interested groups, including health care professionals, politicians, insurers, and the public. In this book we discuss the scientific and technological challenges to the extent necessary to give full attention to the social challenges raised by nutritional genomics. It is not a work of science but rather, a study of the social implications raised by the advent of nutritional genomics. Consequently, we will not comment on where we think efforts in scientific and technological development ought to be directed, since that is not our concern and is out of our range of expertise. We will, however, examine a number of alternatives and make several recommendations about how the social implications of nutritional genomics ought to be addressed.

1.2 WHAT IS NUTRITIONAL GENOMICS?

Our food contains thousands of different compounds, and approximately 30,000 genes constantly rebuild and regulate our bodies. *Nutritional genomics can be broadly understood to involve the study of how nutrients in food interact with our genes at the molecular and cellular levels, and the impacts these reactions have on our health.* Nutritional genomics has two main goals:

1. To understand the functional interaction between bioactive food components with the genome at the molecular, cellular, and systemic level in order to understand the role of nutrients in gene expression and more important, how diet can be used to prevent or treat disease,
2. To understand the effect of genetic variation on the interaction between diet and disease, focusing on each person's specific response to food, due to genetic variants or polymorphisms so as one day to develop dietary recommendations regarding the risks and benefits of specific diets or dietary components to individuals as well as to populations.

Ultimately, the hope is that scientifically sound information on nutrient–gene interactions will lead to strategies that might prevent, mitigate, or aid in the treatment of diseases such as cancer and cardiovascular disease.

In a new field such as nutritional genomics, issues of terminology are not always settled. Nutritional genomics, also known as *nutrigenomics*, is the integrative science at the interface of nutrition, molecular biology, and genomics. *Nutrigenomics* will continue to be the dominant term for this field, but it is important to recognize the meanings of related terms (Ordovas and Mooser 2004). As we discuss in greater detail in Chapter 2, it is sometimes the case that the terms *nutrigenomics* and *nutrigenetics* are used. When these terms are contrasted in this way, *nutrigenomics* refers specifically to the role of nutrients in gene expression (approximating to the first goal above), whereas *nutrigenetics* refers to how genetic variants or polymorphisms can affect responses to nutrients (approximating to the second goal above). Although it can be confusing, since *nutrigenomics* can also be the shorthand for *nutritional genomics*, the common use of *nutrigenomics* to refer to personalized or individualized nutrition is not problematic. Only in Chapter 2 do we make special use of *nutrigenomics* and *nutrigenetics*, for reasons we explain there.

Nutrigenomics is a recent name given to unify a rapidly developing field of study that has been under way for several years, albeit in a fragmented state. Physicians have long used family histories—patterns of inheritance— to help them determine disease susceptibility for a number of conditions. Pedigree studies have also been important in demonstrating that a condition has a genetic basis. The first genetic test used widely for detection of the

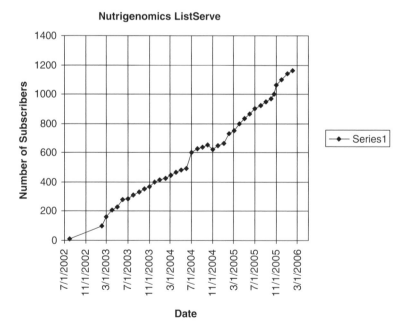

Figure 1.1. The University of California–Davis Nutrigenomics ListServe now has over 1000 subscribers from more than 40 countries (University of California–Davis 2006).

metabolic consequences of genetic disorders was developed in 1963 (Guthrie and Susi 1963) and recommended officially by the American Academy of Pediatrics for widespread implementation in 1965 (American Academy of Pediatrics 1965). It is still used globally to determine whether a newborn suffers from phenylketonuria (PKU), an inherited disease that can cause brain damage unless prevented by a special diet. Contemporary genetic tests can be defined broadly and include DNA analyses to detect heritable disease and the characterization of phenotypes for clinical purposes (Kroese et al. 2004). More important, gene tests can seek to predict the susceptibility to, or probability of, developing a disease in the future, as well as to help diagnose the clinical signs of a disease.

A test for PKU is part of a diagnosis of the disease. Genetic testing in nutrigenomics, by contrast, would seek to predict the future risk of developing diseases in otherwise healthy people, based on the interaction between nutrients and their genes. Unlike a diagnostic test, however, a predictive test does not guarantee the development of disease. Nutrigenetic tests would be used to determine whether a person has a predisposition to a given disease or condition. Once susceptibility is determined on the basis of a person's genotype, the information can be used to suggest dietary interventions tailored to that person's individual genetic profile. For a complex disease such as cardiovascular disease, genetic variants do not tell deterministically if disease will develop, but rather, indicate that there is an increased "risk that a disease will develop in the presence of certain behaviors" or if the variant is protective, "minimizing harmful influences and reducing our risk of disease" (DeBusk and Joffe 2006). The implications are tremendous: Dietary patterns are known to be strongly linked to a number of the greatest causes of illness and death in North America, and among the conditions being studied for gene–diet–health links are hypertension, heart disease, several cancers, diabetes, Alzheimer's disease, schizophrenia, osteoporosis, obesity, glaucoma, and immune system response.

At this time, nutrigenomics is not in widespread use by the public. It has not made its way into public health systems, except in cases in which primary health care practitioners out-source private company tests: for example, Sciona's tests in the United Kingdom and GeneCare's tests in South Africa. At present, there are only about a dozen companies in the world offering nutrigenomic tests. As the science of nutrigenomics advances, commercial interests are expected to grow. One researcher predicts that in as little as 10 years' time, genetic testing will allow us to test for a person's complete set of disease susceptibilities (Peregrin 2001). Althrough this may be as optimistic as former declarations about people leaving the hospital with their genomes encoded on a CD or DVD (BBC 2002), there are, as we discuss below, strong indications that the field will continue to expand.

1.3 METHODOLOGY AND APPROACH OF THIS BOOK

In this book we focus on the social issues—predominantly ethical and legal issues—generated by the development of nutrigenomics science and technology. Our conviction is that the success of a new field such as nutrigenomics depends only partially on the legitimacy of the science and beneficial impact of its applications. The other part is the condition of the social environment in which the science and technology debuts. As the genetically modified food debate between the United States and Europe suggests, it is wholly possible for a technology to be stopped for ethical reasons that may be anchored in different cultural attitudes toward new technology. This is not to suggest that nutrigenomics will go the way of genetically modified foods, but it is to emphasize that good science and technology is only half the way to public acceptance. We also believe strongly that this social analysis needs to occur early, as the field is evolving, and that it ought to evolve along with the science.

Our methodology is to try to understand the current state of nutrigenomics science, together with potential or actual applications, in order to forecast the social issues. This strategy means that our bioethical stance is proactive: As the science and technology progresses, the accompanying ethical and legal analysis should be one step ahead. In this respect, our work differs from what has been described as "bioethics as usual" (*Nature Genetics* 2001). Instead, our proactive stance is meant to address the perceived passivity of much bioethics work by offering timely analysis that can be used to guide ethical science and technology development, and to provide industry and regulators with analyses and recommendations that will guide good business practices and furnish the ground for appropriate regulation.

One of the chief consequences of this methodology is that it can give the appearance that we concentrate wholly on problems that can arise in science and technology. One could object that we do not pay sufficient heed to the benefits offered by a field such as nutrigenomics, or that our work can raise enough issues to dampen enthusiasm about the field's potential to bring meaningful benefits to people. Let us say explicitly and emphatically that this is not the case. In fact, just the opposite is true: Were it not for the fact that we anticipate the potential of nutrigenomics to better people's lives, it would be an unlikely candidate for our proactive approach to studying emerging science and technology. Consequently, throughout this book our approach is to be *conditionally facilitative*, a technically precise way of saying that *if* nutrigenomics can deliver the benefits that it promises, *then* we think certain issues need to addressed. The recommendations we make throughout the book are based on this pattern.

The research on which this book is based derived from the Canadian Program on Genomics and Global Health, funded by Genome Canada through the Ontario Genomics Institute and co-funders identified at the Program's website (Joint Centre for Bioethics 2006). We had the benefit of an expert panel, which provided expert advice at every step in the process of researching this topic and deriving our conclusions. Midway through the process, in 2004, we undertook a limited public consultation by disseminating our preliminary thoughts in a report, *Nutrition and Genes*, and media release, and by seeking feedback. Some of these viewpoints are included throughout the book. We also conducted focus groups with various interested groups, primarily in North America and the UK. The advice we received from the panel of experts, and the feedback received in response to our preliminary report and through the focus groups, has, we believe, made our conclusions in this book more robust. We are grateful to all those who provided input, but, of course, take full responsibility for our conclusions here.

Expert Advisory Panel

George Gaskell
Professor, Department of Social Psychology, London School of Economics

Peter Gillies
Cross-Appointed Professor, Department of Nutritional Sciences, University of Toronto

Jean Jones
Chair, Health Council, Consumers' Association of Canada

Bartha Knoppers
Professor, Faculty of Law, Université de Montreal

John Milner
Chief, Nutritional Science Research Group, National Cancer Institute

Barbara Schneeman
Professor, Nutrition Department, University of California, Davis

Patrick Terry
President, International Genetic Alliance

Stakeholder Dialogues on Personalized Nutrition

The Ethics and Nutritional Genomics Project collaborated with the International Food Information Council, a U.S.-based organization that communicates science-based information on food safety and nutrition, to seek outside opinions on this review. Two meetings with stakeholders were held in Washington in January 2004.

Summary of main Issues from Stakeholder Groups

- A thorough scientific review of nutrigenomic research, especially nutrigenetic tests (i.e., clinical validity/utility) is needed before predictive tests in this area can be commercialized.
- There is a clear distinction between two areas in this field of science: (1) nutrigenomic research and (2) nutrigenetic tests. The report focuses primarily on the ethical implications of area 2. This exclusive focus threatens to undermine the legitimacy of longer-term nutrigenomic research.
- This field is prone to a conflict of interest between commercial and scientific research incentives.
- Currently, nutrigenetic tests are being commercialized prematurely.
- The most ideal method of service delivery for nutrigenetic tests is a team model (i.e., dieticians and M.D. geneticists).
- There is a lack of evidence as to whether personalized nutrition increases motivation to change diet behavior.
- Nutrigenomics could lead to negative impacts on the cultural–social aspect of food.
- Attempting a global health initiative in the area of nutrigenomics may affect the credibility of the report for reasons noted previously.
- Nutrigenomics will enhance consumer–patient empowerment.
- Nutrigenomics will aid in the prevention of modern chronic diseases such as cardiovascular disease.

1.4 OPPORTUNITIES AND CHALLENGES FOR NUTRIGENOMICS

In this section we summarize and discuss some of the benefits we think nutrigenomics may offer and then consider the challenges that lie on the road ahead. A full discussion of these challenges is the main focus of the book.

1.4.1 Improved Health

The greatest benefit expected to flow from nutrigenomics centers on health and quality-of-life improvements. These will be achieved through personalized dietary advice, customized dietary guidelines, improved dietary habits,

Participating Organizations at Washington, DC Nutrigenomics Focus Group, January 2004

American College of Nutrition

American Dietetic Association

Academy of Managed Care Pharmacy

American Nurses Association

American Society for Nutritional Sciences

Division of Nutrition Research
Department of Health and Human Services
National Institutes of Health

Dupont Nutrition and Health

Division of Cancer Research
National Cancer Institute
National Institutes of Health

Office of Dietary Supplements
National Institutes of Health

International Food Information Council

National Coalition for Health Professional Education in Genetics

National Society of Genetic Counselors

National Association of Women, Infants and Children

Office of Disease Prevention and Health Promotion
Department of Health and Human Services

Society for Nutrition Education

Diet and Human Performance Laboratory
Agricultural Research Service
U.S. Department of Agriculture

and the development of healthier foods. The use of a genetically customized diet should not be seen as a rapid "cure" for diseases in the way that an antibiotic can kill bacteria, nor is a customized diet guaranteed to prevent future disease. Adjustments to diets over a long period will probably reduce the risks of a number of diseases and conditions and help people to maintain optimal health. As a result, it may be the relatively young who will benefit most, because their health over a long period may be helped by a tailored diet, particularly if they have genetic susceptibilities to certain illnesses. At present it is less clear how useful nutritional changes based on genetic tests will be in treating existing illnesses.

As we discuss in Chapter 4, nutrigenomics may play an important role in moving health care toward a more proactive or preventive approach. One researcher has suggested that as our knowledge increases, we may be able to test children early in life so that optimal dietary habits, along with other lifestyle patterns, could be fostered before environment-related health damage begins (Fogg-Johnson and Merolli 2000). To the extent that the medical community adopts nutrigenomics as a tool for patients, it may also play a role in more closely linking medicine and nutritional science.

1.4.2   Personalized Dietary Advice

Research on the interaction between genes, diet, and subsequent health outcomes is expected to lead to improved dietary advice tailored to a person's genetic makeup. Current studies about the relationship between diet and health outcomes tend to be based on large population samples. Studies done with large numbers of people do not necessarily lead to accurate predictions about individual responses to dietary components, because of the genetic differences that exist among individuals. Nevertheless, associations between certain foods and increased or decreased disease susceptibility have typically been based on averages from observational studies of large numbers of people. Even when these associations appear to be strong at the population level, the resulting dietary recommendations can be much less precise for an individual. The result is that at present many people are doing little more than making educated guesses about their particular nutritional needs. Nutritional advice that is good for the majority of people can be of no value or may even be harmful for a minority of people with different genetics. There is a subset of people at very high risk for cardiovascular disease whose risks are exacerbated if they follow some of the current dietary recommendations. For example, moderate alcohol consumption is believed to reduce the risk of heart disease, and U.S. guidelines reflect this. But for people with the ApoE4 gene polymorphism, alcohol consumption can raise the level of bad cholesterol, increasing their risk (Corella et al. 2001). Thus, to decrease the risk of bad cholesterol, these people are encouraged to be cautious about their alcohol consumption. With individual genetic profiling, it is possible to engage in what might be called "intelligent" nutrition advice. This will allow more precisely targeted prevention and intervention measures that are likely to reduce the risk of disease and to optimize health.

1.4.3　Improved Diet

There is mixed evidence as to the effect of providing people with nutritional advice. One of the greatest challenges for registered dietitians and allied health practitioners is to get people to modify their behaviors to improve their health. It is not certain that providing people with information about health-related genetic risks will increase their motivation to change behavior more than if they received nongenetic information (Pinsky et al. 2001). However, a survey published in 2000 found that almost 93 percent of Canadians felt that genetic testing would be acceptable if used to predict future illnesses (Martin 2000). The hope is that if nutritional advice is tailored to the individual, there is more reason for the consumer to believe that it will work and therefore to follow the advice. Much nutritional advice can be simple, convenient, easily accessible, relatively inexpensive, and requires little in the way of lifestyle adjustment or special skills. For example, insufficient folate is a known health risk factor in humans. It is now established that when a pregnant woman has low blood folate levels, she is at increased risk of giving birth to a child with neural tube defects, a condition that carries high mortality and morbidity rates for the child. Folic acid supplements have been shown to decrease the rate of neural tube defects in newborns and are now widely used (Medical Research Council 1991).

Not surprisingly, the best audience for new information is among the large number of people already actively seeking health information. Surveys have found that such people are much more likely to change their dietary behaviors than are there less informed counterparts (Institute for the Future 2001). As we discuss in Chapters 4 and 5, health information seekers and early adopters of new health technology and advice are also likely to be among the first users of nutrigenomic services. Because only a few companies offer genetic testing, and have only begun to do so in recent years, it is too early to draw any conclusions about the impact of genetic testing on nutritional compliance rates.

1.4.4   More Development of Health-Enhancing Food Products

Consumers are increasingly looking to food not only for basic nutrition, but also for specific health benefits. This has led to the development of new markets for foods that claim to have specific health effects. New categories of foods, such as functional foods, nutraceuticals, and medical foods, represent one of the fastest-growing markets for the food industry.

- **Functional food**: a food demonstrated to deliver a health-promoting effect beyond basic nutrition.
- **Nutraceutical**: any substance derived and isolated from a food that provides health benefits, including the ability to protect against a chronic disease.
- **Medical or medicinal food**: a food designed for dietary management of a disease or condition that is closely linked to specific nutritional requirements.

Continuing advances in nutrigenomics has the potential to support the development of these new markets both by increasing scientific understanding of the effects of nutrients at the molecular level, and by helping to create consumer demand for these products. In Chapter 5 we consider the impact that the crossover between foods and drugs has for the advancement of nutrigenomics, and how this might be regulated as these emerging markets are driven by consumer demand for healthier food and food ingredients. This demand is expected to rise as baby boomers continue to age and as consumers generally become better educated about their own health and dietary needs.

1.4.5   Consumer Empowerment

With the rise of the patients' rights movement, beginning in North America in the 1960s, there has been a marked shift in the practitioner–patient relationship. It has evolved from a paternalistic model to a consumer model that decreases practitioner dominance and increases patient choice and control (Emanuel and Emanuel 1992). This progression in the role of the individual in health care manifests itself in a number of ways, including the increasing availability and widespread demand for consumer health information, and greater use of traditional, complementary, and alternative medicines (CAMs). Recent studies indicate that approximately 50 percent of people in many industrialized countries use CAMs (Bodeker and Kronenberg 2002). Among the reasons that people give for turning to CAMs is a feeling that the latter are more prevention-oriented than is the case for Western medicine. Some people prefer CAMs because they typically require more active consumer involvement in investigating, protecting, and promoting their health. In developing countries, where CAMs have always been used, up to 80 percent of populations are involved.

The increasing demand for consumer health information helps people to understand their health issues and to make better-informed decisions for themselves and their families (Bouhaddou et al. 1998). This has resulted in increasingly sophisticated consumers who are well educated about health and nutrition. Studies indicate that 85 percent of consumers view food as a key component in maintaining health and managing disease, and according to one study, the majority of such consumers consider food labels to be one of the most valuable sources of nutrition information (Peregrin 2001). In Chapter 4 we discuss how nutrigenomics is likely to become part of this trend toward increasing consumer empowerment by providing additional tools and information to those consumers who engage in self-care by gathering information and choosing remedies. In particular, people will have the option of using dietary modification rather than pharmaceuticals for conditions such as high cholesterol.

1.4.6 Reducing Health Disparities

In Chapter 6 we turn our attention to how nutrigenomics may have a role to play in reducing health disparities between social groups. Although most genetic variation is shared across racial and ethnic groups, in some cases there are medically important differences between groups, including susceptibility to certain diseases. Some populations can have disproportionately high incidences and severity of such chronic diseases as diabetes, asthma, cardiovascular disease, and certain cancers. For example, African-American men have a 60 percent greater risk of developing prostate cancer and are two to three times more likely to die from it than are Caucasian men. A number of factors are thought to contribute to these disparities, including the interplay between genetics and diet. Genomics is making it possible to analyze the genetic differences that may underlie disparate group rates in the incidence or patterns of progression for these diseases. In addition to addressing such social factors as inequities in access to health care and living in poor environments, the information could be used to help reduce health disparities. For example, this research may lead to community outreach programs to inform affected minorities about the importance of specific nutritional habits related to their particular genetic makeup. A National Center of Excellence in Nutritional Genomics, supported by the U.S. National Institutes of Health, was established at the University of California–Davis to study such issues (University of California–Davis 2006).

1.4.7   Health Care Savings

One potential benefit of nutrigenomics is cost savings for consumers, employers, governments, and third-party payers (such as insurance companies) through the prevention and delay of disease onset. In Europe, for example, governments spend approximately 5 percent of their national health care budgets on obesity-related health problems (World Health Organization 2000). In addition to these direct costs, there are indirect costs, such as loss of ability to work due to early disease and deaths. With obesity becoming a major health crisis in a growing number of countries, nutrigenomics research designed to identify genetic predispositions to this problem could lead to significant health care savings. There are similar hopes for fighting cancer. It has been proposed that practical dietary changes could result in a 35 percent reduction in cancer rates in the United States (Elliott and Ong 2002). Nutrigenomics may lead to nutrition recommendations tailored to reflect more accurately the diversity in response to food among individuals and subpopulations. It could lead to the eventual modification of population-based food guidelines. The other longer-range hope would be to find bioactive components of food matched to genotypes that could be used in treatments or protection using food or dietary supplements instead of generally more expensive drugs. Such savings would benefit individuals, governments, and third-party payers who provide drug coverage.

1.5   CHALLENGES AND A ROAD MAP OF THIS BOOK

Nutrigenomics faces a number of challenges. Despite great promise, nutrigenomics is an evolving field, and it is still difficult to quantify the many benefits. Researchers are only able consistently to replicate a small number of gene–disease associations. It appears that in most cases, multiple genes, as well as environmental factors, including foods, are involved in the development of such complex illnesses as cancer and cardiovascular disease. Human biology is complex, and it will take time and large, expensive population studies to get greater insights into the interactions between genes and nutrition. As we discuss in Chapter 2, an accumulation of uncertainties, methodological differences, and margins of error in nutrigenomics studies can make it difficult to reach scientific consensus. Consequently, one might wonder whether enough good-quality science is available to make personalized dietary recommendations based on genes. As one commentator puts it, "there is not even the information needed for setting dietary recommendations with confidence now at the group level" (Arab 2004), a fact attributed to the still limited ability of current tools to study the complex interrelationships among diet, genetics, and environment.

Nutrigenomics requires the collection of biological samples from which genetic information is extracted. Like any other genetic tests, tests for links between genes and food require a rigorous evaluation of the social implications, including ethical, economic, and legal concerns they may raise. In Chapter 3 we consider questions such as who ought to have access to biological samples or customers' genetic profiles and how those samples and profiles will be used in relation to insurance and/or employment. The potential for accidental or deliberate release of personal genetic information gathered during research and in the provision of clinical or private nutrigenomic services raises concerns about the need for appropriate standards regarding privacy and confidentiality.

It is already the case that some nutrigenomic services are being provided privately, and these have drawn the attention of regulators and the media. There are unresolved questions about how nutrigenomic services ought to be provided to the public, and in Chapter 4 we consider a number of potential service delivery models. We examine such issues as whether nutrigenomics should be translated into clinical contexts as opposed to being offered to consumers directly. Primary health care practitioners will have to be willing to adopt and use the new information as well as be competent to use the relevant knowledge, and to administer and/or interpret nutrigenomic, tests, safely and effectively. Accordingly, when it comes to providing nutrigenomic services, there are a number of challenges, including lack of trained personnel, inadequate resources, and questions of reimbursement for a new service that may be seen as displacing other services that are currently funded.

Regulating food–health claims is already difficult, and tests that promise to help a person choose a diet based on a genetic profile add one more layer of complexity. Furthermore, nutrigenomics sometimes seems to straddle the ground between foods and medicines, which are traditionally regulated separately. This poses problems for regulating the tests and the products as regulators face new challenges in how they will regulate foods that claim health benefits approaching the realm of drugs (Chadwick 2004). It may be the case that regulatory structures, approaches, and definitions will have to adapt to fields such as nutrigenomics, rather than the other way around (Castle et al. 2006). In Chapter 5 we consider some of these regulatory reforms and innovations in detail in light of how nutrigenomics may cross traditional boundaries between foods and drugs and accelerate the development of supplements, functional foods, and nutraceuticals.

Questions of equity and access to this expensive new science both within and among nations are the subject of Chapter 6. Nutrigenomics is not currently publicly funded, and it is available only to a small population that can afford such services. In developing nations, access to basic food and health care are priorities, so there is a challenge in how to make the emerging technologies available to a wide range of people if the potential for widespread benefits of nutrigenomics are to reach as many people as possible. In this context, the role that intellectual property rights play in limiting access to new technologies is discussed.

In each chapter we offer some preliminary conclusions and recommendations in relation to the issue discussed. We summarize these in the final chapter. In light of the opportunities and challenges that have been identified, we believe that nutrigenomics can provide new opportunities for disease prevention and health promotion, This new field also raises ethical, legal, and social issues about which we provide analysis, draw conclusions, and make recommendations. We encourage an ongoing public discussion regarding the latest knowledge of the opportunities and limitations of nutrigenomics in improving health. This discussion should foster greater public awareness and understanding of nutrigenomic concepts and applications.

REFERENCES

American Academy of Pediatrics, Committee on the Fetus and Newborn. 1965. Screening of newborn infants for metabolic disease. *Pediatrics* 35:499–501.

Arab, L. 2004. Individualized nutrition recommendations: do we have the measurements needed to assess risk and make dietary recommendations? *Proceedings of the Nutrition Society* 63:167–172.

BBC. 2002. Your genetic code on a disc. Retrieved September 23, 2002, from http://bbc.co.uk/2/hi/sci/tech/2276095.stm.

Bodeker, G., and F. Kronenberg. 2002. A public health agenda for traditional, complementary and alternative medicine. *American Journal of Public Health* 92:1582–1592.

Bouhaddou, O., J. G. Lambert, and S. Miller. 1998. Consumer health informatics: knowledge engineering and evaluation of studies of Medical HouseCall. Presented at the American Medical Informatics Association Annual Symposium.

Castle, D., R. Loeppky, and M. Saner. 2006. Convergence in biotechnology innovation: case studies and implications for regulation. University of Guedph. Retrieved from www.gels.ca.

Chadwick, R. 2004. Nutrigenomics, individualism and public health. *Proceedings of the Nutrition Society* 63:161–166.

Corella, D., K. Tucker, C. Lahoz, O. Coltell, L. A. Cupples, P. W. Wilson, E. J. Schaefer, and J. M. Ordovas. 2001. Alcohol drinking determines the effect of the ApoE locus on LDL-cholesterol in men: the Framingham Offspring Study. *American Journal of Clinical Nutrition* 73:736–745.

DeBusk, R., and Y. Joffe. 2006. *It's Not Just Your Genes*. San Diego, CA: BKDR Publishing.

Elliott, R., and T. J. Ong. 2002. Nutritional genomics. *British Medical Journal* 324(7351):1438–1442.

Emanuel, E. J., and L. L. Emanuel. 1992. Four models of the physician–patient relationship. *Journal of the American Medical Association* 267:2221–2226.

Fogg-Johnson, N., and A. Merolli. 2000. Nutrigenomics: the next wave in nutrition research. Retrieved February 26, 2006, from www.nutraceuticalsworld.com/marapr001.htm.

Guthrie, R., and S. Susi. 1963. A simple phenylalanine method for detecting phenylketonuria in large populations of newborn infants. *Pediatrics* 32:338–343.

Hasler, Clare M. 2000. The changing face of functional foods. *Journal of the American College of Nutrition* 19(5):499S–506S.

Institute for the Future. 2001. *The Future of Nutrition: Consumers Engage with Science.* Cupertino, CA: IF.

Joint Centre for Bioethics. 2006. Canadian Program on Genomics and Global Health. Retrieved February 29, 2006, from www.geneticsethics.net.

Kroese, M., R. L. Zimmern, and S. Sanderson. 2004. Genetic tests and their evaluation: Can we answer the key questions? *Genetics in Medicine* 6:475–480.

Martin, S. 2000. Most Canadians welcome genetic testing. *Canadian Medical Association Journal* 163:200.

Medical Research Council Vitamin Study Research Group. 1991. Prevention of neural tube defects: results of the Medical Research Council Vitamin Study. *Lancet* 338:131–137.

Mendelson, R., V. Tarasuk, J. Chapell, H. Brown, and G. H. Anderson. 2003. Report of the Ontario Food Survey. Toronto: Province of Ontario.

Milner, J. 2002. Functional foods and health: a US perspective. *British Journal of Nutrition* 88:S151.

Nature: Genetics. 2001. Defining a new bioethic. *Nature: Genetics* 28:297–298.

Ordovas, Jose M., and Vincent Mooser. 2004. Nutrigenomics and nutrigenetics. *Current Opinion in Lipidology* 15:101–108.

Peregrin, T. 2001. The new frontier of nutritional science: nutrigenomics. *Journal of the American Dietetic Association* 101:1306.

Pinsky, L., R. Pagon, and B. Wylie. 2001. Genetics through a primary care lens. *Western Journal of Medicine* 175:47–50.

Rowe, S., and C. Toner. 2003. Dietary supplement use in women: the role of the media. *Journal of Nutrition* 133:2008S–2009S.

Simopoulos, A. 2004. Genetic variation: nutritional implications. In *Nutrigenomics and Nutrigenetics*, edited by A. Simopoulos and J. M. Ordovas. Basel, Switzerland: Karger.

Sloan, A. E. 2000. The top ten functional food trends. *Food Technology* 54:54.

Troppmann, L., T. Johns, and K. Gray-Donald. 2002. Natural health product use in Canada. *Canadian Journal of Public Health* 93:426–430.

University of California–Davis. 2006. The NCMHD Center of Excellence for Nutritional Genomics. Retrieved February 27, 2006, from http://nutrigenomics.ucdavis.edu.

Verschuren, P. M. 2002. Functional foods: scientific and global perspectives. *British Journal of Nutrition* 88(S125).

World Health Organization. 2000. *The World Health Report 2000—Health Systems: Improving Performance*. Geneva, Switzerland, WHO.

2

THE SCIENCE OF NUTRIGENOMICS AND NUTRIGENETICS

2.1 INTRODUCTION

Nutrition is just one of many environmental factors that play a role in the development of disease and maintenance of health. Other environmental factors can be broadly defined to include infectious, chemical, physical, and even behavioral factors. Nutrition is, however, the single environmental factor to which all humans are permanently and continuously exposed from conception to death (Ordovas et al. 2004). Its significance in how we understand human health and disease is emphasized by the fact that poor dietary habits are responsible for more than half of the leading causes of death in the United States (Davis and Milner 2004). The global prevalence of chronic diseases such as cardiovascular disease, diabetes, cancer, and chronic respiratory disease is growing exponentially, and the role of diet is at the forefront in understanding and potentially preventing such increases (Yach et al. 2006). As a result, a growing number of scientists advocate, and are developing, integrative approaches to the study of human health and disease in relation to our dietary habits.

Nutritional genomics, also commonly referred to as *nutrigenomics*, is the integrative science at the interface of nutrition, molecular biology, and

Science, Society, and the Supermarket: The Opportunities and Challenges of Nutrigenomics,
By David Castle, Cheryl Cline, Abdallah S. Daar, Charoula Tsamis, and Peter A. Singer
Copyright © 2007 John Wiley & Sons, Inc.

Nomenclature: Nutritional genomics = Nutrigenomics

Nutrigenomics is the term most commonly used for this field. It is sometimes broken down into two subfields to contrast the difference between genomics and genetics:

Nutritional genomics, a.k.a. *nutrigenomics:* the integrative science at the interface of nutrition, molecular biology, and genomics.

{

Nutri*genomics* investigates the role of nutrients in gene expression.

Nutri*genetics* is the study of how genetic variants or polymorphisms can affect responses to nutrients.

In this chapter we differentiate between nutri*genomics* and nutri*genetics* in order to capture their differentiating applications and goals. Throughout the rest of the book we use the term *nutrigenomics* to capture both senses of the term.

---

genomics. In a way similar to the uses of pharmacogenomics and pharmacogenetics, a distinction can be made between nutrigenomics and nutrigenetics. Technically speaking, nutritional genomics integrates two subspecializations: *nutrigenomics* and *nutrigenetics*, although in common use the term *nutrigenomics* refers to both the genomic and genetic aspects of the field. Whereas nutrigenomics investigates the role of nutrients in gene expression, nutrigenetics is the study of how genetic variants or polymorphisms can affect responses to nutrients.

It is generally the case that people will use the term *nutrigenomics* intending it to refer to both genomic and genetic studies of diet–gene interaction. In the nutritional genomics literature, it is sometimes appropriate to highlight the genomic aspects of the field, in which case nutrigenomics will sometimes be written as nutri*genomics* to contrast it with nutri*genetics*. In this chapter we find occasion to distinguish between the two technical definitions for nutrigenomics and nutrigenetics in order to capture their differentiating applications and goals. Throughout the rest of the book we use the term *nutrigenomics* to capture both senses of the term.

Nutri*genomics* examines the functional interaction between bioactive food components with the genome at the molecular, cellular, and systemic level, and its main goal is to understand the role of nutrients in gene expression, and more important, how diet can be used to prevent or treat disease. Nutri-

genetics can be understood as a subfield of nutritional genomics since it examines the effect of genetic variation on the interaction between diet and disease and focuses on an individual's specific response to food due to genetic variants or polymorphisms. Nutri*genetics* has also been termed *personalized nutrition* or *individualized nutrition* because its main goal is one day to develop dietary recommendations regarding the risks and benefits to individuals of specific diets or dietary components. Propelled by this individualized approach to health is the commercialization of genetic tests geared toward offering customized dietary advice to consumers.

The integration of these several disciplines is paving the way for the understanding and identification of genotypes that are risk factors for the development of diet-related diseases. Both nutrigenomics and nutrigenetics are anticipated to lead to significant improvements in our health by contributing to an ongoing revision of general nutrition guidelines, and through the development of personalized diets to achieve optimal health and to reduce the risk of a variety of diseases. Nutrigenomics and nutrigenetics are also said to affect the food industry through the development of novel foods and substantiated health claims, the implications of which we discuss in detail in Chapter 5.

Before these goals can be achieved, nutrigenomics and nutrigenetics face a number of challenges. First, more information is required that links specific nutrient–gene interactions with functional outcomes. Additional well-designed and conducted epidemiological studies are needed, and their application to the study of nutrigenomics and nutrigenetics needs further elaboration. More important, scientists face the challenge of translating nutrigenomics and nutrigenetics information into accurate and effective predictions of the beneficial and adverse health effects of dietary components in relation to chronic diseases. In addition to developing a robust science of nutrient–gene interactions, the science must have proven technological applications as well as health benefits. The commercial and clinical application of nutrigenomics and nutrigenetics information must be demonstrably efficacious and safe, or must at least meet some standard of tolerable risk. An additional question, which we consider in Chapter 5, is whether nutrigenetic interventions must be effective as current pharmaceutical interventions and be regulated that way, or whether the interventions are more akin to the individual choice of regulated foods and food products.

Our approach to this review of nutrigenomics and nutrigenetics will be guided by a single question about the field as a whole: Is the science of nutrigenomics sufficiently strong to warrant claims of health benefits? To answer this question we first locate nutrigenomics and nutrigenetics in their broader scientific contexts before examining some of the strengths and weaknesses of this integrative science. We then turn to the most important regulatory and ethical consideration: the support nutrigenomics and nutrigenetics offers for claims that some dietary changes may offer health benefits. What is not attempted here is a comprehensive literature review of the science—we leave that to the relevant scientific communities. Rather, we consider key aspects of the science that bear directly on the ethical and regulatory need for a clear understanding of the potential benefits and risks. A fair assessment of the science and its applications will be essential to the development of appropriate regulations and ethical guidelines for this emerging field.

2.2   THE SCIENTIFIC CONTEXT

The dynamic interaction between genetic and environmental factors has been proposed for several decades now to study the effects of genetic variation and gene–nutrient interactions in the management of chronic diseases (Holtzman 1988; Simopoulos et al. 1990) For example, the term *nutrigenetics* was first used and defined by R.O. Brennan in 1975 in his book *Nutrigenetics: New Concepts for Relieving Hypoglycemia* (Brennan 1975). Although the existence of a genetic component has been acknowledged for several decades, only recently with the aid of high-throughput and/or genomic technologies have we begun to understand the complex and dynamic interactions between nutrients and genes and their role in the onset and progression of chronic diseases. Patrick Stover claims that "nutrition has been perhaps the most persistent and variable of the environmental exposures that have challenged and thereby shaped the human genome and contributed to its variation[3] (Stover and Garza 2002)." The reciprocal or bidirectional interaction between genes and nutrients is illustrated by the fact that on the one hand, nutritional components can influence gene expression, and on the other, genetic variations can affect our response to nutrients. These interactions are then linked to the maintenance of health or the onset and progression of diseases.

Although the first good draft of the human genome (Lander et al. 2001; Venter et al. 2001) has generated an enormous amount of new data potentially useful for health, this information is just beginning to generate ideas about how to maximize health benefits. Despite repeated claims in the scientific literature and the popular press that genomics will revolutionize medicine, there are few obvious current clinical and public health applications directly attributable to the Human Genome Project (HGP) (Wright and Hastie 2001). The benefits of the HGP are often hyperbolized in the media (Bubela and Caulfield 2004). As it stands, the number of applications of the

HGP are few and disproportionate to the vast amount of data collected. Since the HGP cost roughly U.S. $3 billion, the pressure to see the science translate into benefits for the ordinary citizen is understandable.

A more realistic time frame for a surge in clinical applications of human genomics is a decade or two from now (Khoury 2003). The number of genes discovered in the draft of the human genome was roughly a third (ca. 30,000) as many as originally estimated. Many of the paradigmatic gene-based diseases known in medicine, such as Tay–Sachs and Huntington's, are attributable to defects in single genes. A one-gene one-disease paradigm attributes narrow functions to genes and characterizes gene-based disease as simple deviations from normal functioning of single genes. When the consensus emerged that the number of genes in the human genome was far fewer than predicted, the monogenic disease paradigm that dominated a generation of molecular medicine no longer fits all the facts.

The HGP reveals that genes must have multiple functions, they work in concert with one another, and they interact with their environment. Human genomics is the study of our total complement of genes, understood as a complex system interacting in the ways just described. The HGP ushered in the *genomics era*, a term that reflects the momentous conceptual "sea change" that genomics has brought to biology. There are even calls for a new, postgenomic era in recognition of the need to fully integrate the study of genomes, the protein *proteome*, and the resulting metabolism or *metabolome*. Meanwhile, others are turning to a study of the environment's impact on the genetic basis of behavior (Caspi et al. 2002; Ridley 2003), the effect of toxins (Schmidt 2002), the effects of drugs (Evans and Johnson 2001; Evans and McLeod 2003), and of course, nutrigenomics. Study of these complex, integrated systems may ultimately yield a complete understanding of human biology (Hood 2003), but it could come with a price if the complexity of the science prohibits us from achieving the basic goals of improving health (Cooper and Psaty 2003). The sheer complexity of the task at hand has alerted biologists that new methods of classifying phenomena supported by dynamic systems biology models are incremental steps toward understanding and applying human genomics (Jimenez-Sanchez et al. 2001; Cassman 2005).

To summarize so far, an assessment of nutrigenomics and nutrigenetics must take into account the complexity and novelty of the scientific approaches themselves, and a fair assessment of such a field must also recognize that translating human systems biology—itself a field undergoing rapid evolution—into clinical applications is a major undertaking. Nutrigenomics and nutrigenetics only recently became an integrative field of intense international scientific inquiry, as a PubMed search for "nutrigenomics, nutritional genomics, and nutrigenetics" reveals.

The European Nutrigenomics Organization (NuGO) describes itself as "The Network of Excellence in Nutrition and Genomics." The organization (www.nugo.org) started in January 2004 and currently has 22 partner organizations in 10 European countries, backed by EU funding for six years. NuGO is a network dedicated to the integration and facilitation of research, education, communication, commercialization, and dissemination of nutrigenomics in Europe.

Perhaps the ultimate determinant of the value of nutrigenomics and nutrigenetics research will be its ability to deal with well-defined clinical problems and to contribute to the well-being of people around the world. A number of high-throughput technologies are also being used to help determine optimal nutrition at the level of populations, groups, and individuals. They expand our understanding of nutritional adequacy, nutrient–gene interactions, dietary recommendations, and effective strategies for their implementation (Elliot and Ong 2002). Nutrigenomics and nutrigenetics are leading to a new paradigm in nutritional science, one that has been likened to a new frontier (Peregrin 2001), or to crossing the Rubicon (Gillies 2003), and may ultimately soften the traditional division between nutrition and medicine. Much more specific studies on the relationship between nutrients and an individual's genetic makeup are now possible, including genomic profiling, which tests people for nutrient-sensitive gene variants associated with greater risk or predisposition for conditions or diseases. Time will tell if science can deliver information that will enable people to make dietary changes that will improve their health. Not surprisingly for a new field, there is a wide range of scientific opinion about the value of nutrigenomics and nutrigenetics research and their related commercial and clinical applications, ranging from unrestrained optimism about the potential to improve health to skepticism about the validity of the science and its practical usefulness.

2.2.1 Nutrigenomics

The current trend is to study nutrient–gene interactions believed to play a role in the diseases most common among the populations of Western nations.

Diseases that are targeted in the field of nutrigenomics include inflammatory diseases (Kornman et al. 2004), diabetes (Wood 2004), obesity (Middleton et al. 2004), osteoporosis, cardiovascular disease, metabolic syndrome (Roche et al. 2005), and a number of cancers (Davis and Milner 2004). The expectation is that genomics-based nutritional sciences will generate knowledge that can usefully lead to revised dietary guidelines. Nutrition guidelines that are informed by a science of nutrient–gene interactions may bring about significant health outcomes and economic benefits—at least to the extent that the guidelines are followed.

At present, nutrigenomic research holds out great promise for far-reaching improvements in health. One powerful approach to understanding the role of nutrient–gene interactions in health is to conduct nutritional intervention studies. It is possible to modulate the expression of genes, and hence affect entire metabolic pathways, by varying the intake of nutrients. Biomarkers are used to measure the effects of diet on gene expression by tracking variations in the levels of gene products and metabolites. This methodology is used widely to elucidate the interaction between genes and metabolic pathways, but the complex nature of nutrient–gene interactions presents a constant challenge. Jim Kaput and Ray Rodriguez (2004) argue that by sticking to the basic methodologies and tenets of nutrigenomics (i.e., that gene expression is influenced measurably by nutrients and can have long-term effects on disease progression), complex nutrient–gene interactions will be resolvable. In relation to chronic diseases, research is expanding rapidly in the area of the effects of dietary cholesterol and fatty acids on gene expression. Other studies include the variety of actions exerted by amino acids and carbohydrates in controlling the expression of genes involved in various biological systems (Kaput and Rodriguez 2004).

Nutrient intervention studies are important tools of discovery for nutrigenomics, but the other side of the coin is that dietary interventions can be used to prevent, mitigate, or otherwise manage diseases with a suspected or known genetic component. In these cases, dietary changes are made to correct for the effects of a particular genotype, but the nutritional intervention is not intended to alter gene expression. Cases of this kind are well known. In the area of pediatrics, ketogenic diets (diets that are high in fat and low in carbohydrates and protein) are used for the treatment of patients with intractable epilepsy. Diets balanced in essential fatty acids are now known to be important for the management of a number of chronic inflammatory diseases, including arthritis, ulcerative colitis, and lupus, as well as for the prevention and treatment of coronary artery disease and hypertension. In the case of single-gene disorders, the approach can sometimes be simple. For example, single-gene disorders such as galactosemia, celiac disease, familial hypercholesterolemia, lactose intolerance, and PKU can largely be controlled by

---

**Nutrient Intervention to Correct for Final Phenotypes of
Gene-Based Metabolic Derangement**

PKU. Phenylketonuria is caused by an inherited mutation on chromosome 12
that produces a defective form of the enzyme phenylalanine hydroxylase,
which is required for converting phenylalanine to tyrosine. Without this con-
version process, phenylalanine, a common component in milk, eggs, bread,
and meat, can build up in a child's body, leading to tremors, seizures, and
brain damage (Koch 1999). Newborn PKU screening is accomplished by
testing a drop of blood for excessive concentrations of the amino acid phenylalanine. The negative consequences of PKU can be prevented by avoiding
foods that contain phenylalanine.

---

specific dietary interventions and may be considered the "classic" cases of
gene–diet interactions (Ordovas and Corella 2004).

What is revolutionary about nutrigenomics is that nutritional interventions
are not only used to correct metabolic deficiencies, but are now being
exploited for their potential to change gene expression and to intervene in
diet–gene interactions related to multifactorial diseases such as cardiovascular disease, cancer, osteoporosis, and neurological diseases associated with
the aging process. Here, the objective is to study the relative contributions
of different genes and their variable expression due to nutrient metabolism
on the complex processes of disease progression. When the final phenotype
(e.g., cardiovascular disease) is attributable to multiple coextensive causes,
the science is obviously much more complex than in classic cases of
diet–gene interactions. The principle of searching for gene–diet interactions
remains the same but needs to be considered in the context of how other
diet–gene interactions affect overall gene expression and metabolism.
Increased cardiovascular disease risk caused by plasma lipids is a case in
point, since there are many genes (e.g., the suite of apolipoprotein loci,
hepatic lipase) known to interact with plasma lipids with serious health
implications (Ordovas and Corella 2004).

Nutrigenomics research is not without its pitfalls. A lack of consistency
in the reported findings of studies done on some complex disorders has been
pointed out. This lack of consistency may be an inherent error rate in high-
throughput laboratory techniques or is perhaps due to variations in laboratory practice (Ordovas 2003). In other cases, the small sample sizes used
in many studies is a source of concern (Mathew 2001). Overcoming these
uncertainties requires using larger study populations, which imposes greater
expense and longer time delays. The current scientific literature suggests a
need to use large sample sizes and controlled dietary intakes to deal with the
challenges faced because of the complexity of genotypes, diets, and their
interactions (Kaput 2004). A case in point is the British undertaking to

> **Nutrient Intervention to Alter Gene Expression**
>
> *Apolipoprotein E (ApoE).* High cholesterol levels contribute to the buildup of plaque in arteries, greatly increasing the risk of coronary heart disease and high blood pressure. For over 20 years scientists have known that the response of plasma cholesterol concentration to dietary cholesterol varies among individuals (Simopoulos 2002). However, only recently have the mechanisms underlying this variation begun to be uncovered. Recent studies suggest that genetic variants of the apolipoprotein ApoE gene may play a role in these observed differences in dietary response. Researchers have found that when a population is placed on a low-fat diet, roughly half experience an improvement in their serum lipoprotein profiles, thereby reducing their risk for cardiovascular disease, while the other half actually show a deterioration in their serum lipoprotein profiles, potentially placing them at increased risk for cardiovascular disease. This difference in response has been attributed to the differential expression of genetic variants of ApoE.

study 500,000 adults drawn from the general population to explore the interactions between genes and environmental and lifestyle risk factors (www.ukbiobank.ac.uk).

If the technical issues just described might apply equally to other genomics-based sciences, a problem that continues to confound nutrigenomics is the lack of reliable data on people's dietary habits, combined with basic metabolic science. As one researcher has put it, this knowledge is "absolutely essential for the understanding and interpretation of nutrigenomics data and to provide dietary advice to improve health" (Fairweather-Tait 2003). Without an adequate baseline of knowledge about normal metabolic states for comparison, the results from nutrient-intervention studies will not have much meaning.

2.2.2 Nutrigenetics

Whereas nutrigenomics investigates the role of nutrients in gene expression, nutrigenetics is the study of how genetic variants or polymorphisms can affect responses to nutrients. Nutrigenetics can be viewed as the domain of clinical applications of nutrigenomics because it can account for individual genetic variation, differential response to nutrients, and disease history. It also provides a context for the commercialization of nutrigenomics research. If nutrigenetics is to be widely accepted by the medical and public health communities, as well as by those who pay for health care, it will have to prove that it can deliver results.

The most common source of genetic differences comprises genetic variants called *single-nucleotide polymorphisms* (SNPs). An SNP is a single-base

The SNP Consortium Ltd. is a nonprofit foundation organized for the purpose of providing public genomic data (http://snp.cshl.org). Its mission is to develop up to 300,000 SNPs distributed evenly throughout the human genome and to make the information related to these SNPs available to the public without intellectual property restrictions. Single nucleotide polymorphisms (SNPs) are common DNA sequence variations among individuals. They promise to advance significantly our ability to understand and treat human disease. The SNP Consortium (TSC), a public–private collaboration, has to date discovered and characterized nearly 1.8 million SNPs.

mutation in DNA that appears in 1 percent or more of a population. For the most part SNPs are benign but when located at critical points in a gene may affect how the gene is expressed, or may produce structural and functional changes in the products of genes. SNPs are also called *functional polymorphisms*, to differentiate them from the bulk of polymorphisms, which have no phenotypic effect. Having a draft of the human genome has enabled researchers to better identify gene sequence variations, so now the race is on to correlate SNPs with known phenotypes (i.e., diseases, drugs, food responses, etc.). Although there are reasons to doubt the appearance of a hard and fast rule that common SNPs always go together with common diseases (Wright and Hastie 2001), some SNPs are clearly responsible for genetic disease or disease susceptibility. When the activity of genes is governed in whole or in part by the presence of specific nutrients, the presence of SNPs may lead to very different responses to the nutrients. In this way, the presence and effects of certain SNPs explain why, among people who consume the same diet, some fail to acquire certain diseases while others have greater-than-expected disease patterns (Fogg-Johnson and Kaput 2003).

Susceptibility to most common diseases is thought to be the result of not just one gene but the combined action of a number of different genes, each of which confers a moderate degree of risk over a long period of time. In addition, and this applies equally for monogenic and polygenic disorders, the presence of a mutation does not always guarantee that a particular disease will result. Such differences in the effect of genes, referred to as their *penetrance*, make long-term predictions for disease progression all the more uncertain. Complicating matters further, it is also now believed that a single gene can have effects through a number of unrelated metabolic pathways and may be associated with a number of different clinical outcomes. For example, variability at the ApoE location has been associated not only with lipid metabolism and cardiovascular risk, but also with some of the most common age-related conditions, including cancer, neurological disorders, and osteoporosis. Finally, even when a genetic mutation known to increase risk is present, actual outcomes may be influenced not only by interactions with other genes but also with a variety of environmental exposures and behaviors (Burke et al. 2002). Under these different circumstances, candidate gene studies, which seek to find an association between specific gene polymorphisms and markers of disease risk, lack precision and may give spurious results.

Recent dietary guidelines have been implemented to improve the health of the general population and of those at high risk for specific diseases, such as cardiovascular disease, cancer, and diabetes. One of the main objectives of nutrigenetics and nutrigenomics research is to develop dietary interventions based on knowledge of nutritional requirement, nutritional status, and genotype in an effort to prevent, mitigate, and cure chronic diseases. Although these practical applications will be realized some years into the future, the shift toward individualized health is already being integrated in nutritional health domains. Governmental organizations such as the U.S. Department of Agriculture have developed the USDA Healthy Eating Index website to help people tailor individual diet and exercise programs (`www.cnpp.usda.gov.healthyeating.html`) and have recently attempted to individualize the American Food Guide Pyramid (`www.mypyramid.gov/`).

Current dietary guidelines however, have not yet taken into account the differences in individual responses to nutrient intake and the vital role that genetic variation plays in these differences. Furthermore, if the use of genotypes in the dietary mitigation and prevention of disease is to be established, future studies must be improved to provide stronger evidence for claims that dietary interventions will have their intended effects. Studies with much larger sample sizes and carefully controlled dietary interventions are desirable, and these should investigate the effects of polymorphisms in multiple genes, instead of the effects of polymorphisms in single genes, if they are to reflect the complexities of multifactorial diseases (Simopoulos 2002).

But, of course, not all nutrigenetics testing has been aimed at the clinical context, where it would be expected to meet stringent criteria. The market is somewhat more permissive about substantiating health claims than is the clinical context, even if the intent is to have the test ordered and interpreted in a clinical manner. Tests to determine genetic risk profiles are currently being offered by a small number of companies in Canada, the United States, and the UK. These tests promise benefits in terms of both wellness promotion and disease prevention. One company offers a test for genetic variants claimed to be related to free radical damage, detoxification, alcohol metab-

Innovative Testing for Optimal Health

Genova Diagnostics (GDX), previously known as Great Smokies Diagnostic Laboratories (GSDL), was established in 1986 in Asheville, North Carolina, to provide "laboratory functional testing" that assesses the "inter-relationship of physiological systems." GDX (www.gdx.net) offers 125 assessments in areas of digestion, nutrition, detoxification–oxidative stress, immunology, allergy, and hormone and metabolic function. They developed *predictive genomics* in 2002, which reveals single-point mutations (SNPs), which can impair the production of certain proteins and enzymes, and the new genomic line was called Genovations (www.genovations.com). Initially, it measured the Cardio Genomic Profile (heart), the Osteo Genomic Profile (bones), and the Immuno Genomic Profile (immune system). Genovations says that its genetic tests identify risks but do not diagnose disease or illness. Tests are ordered for patients by physicians, and the results can help a physician give advice on lifestyle adjustments, including changes in diet.

olism, and skin and hair repair. This is coupled with nutritional advice designed to help reduce identified susceptibilities. One company developed tests for genetic variants purportedly associated with cardiovascular disease, obesity, osteoporosis, a number of detoxification defects related to increased risk for certain cancers, chronic fatigue, multiple chemical sensitivity and alcoholism, and immune system defects which increase susceptibility to asthma, atopy, osteopenia, heart disease, and infectious diseases. Another offers genetic tests that it claims will help people determine which nutritional supplements or skin careproducts are most suited to them on the basis of their individual genotypes.

2.3   THE CASE OF MTHFR

Having discussed each of the two streams, nutrigenomics and nutrigenetics, it is fair to say that this new field is a rich combination of noted promise and promissory notes. Pioneers of this research have unquestionably opened a new field, and there is now strong evidence suggesting that nutrient–gene interactions are associated with long-term health outcomes. As with any new field of study, however, much labor must be accomplished in the trenches. Some of the empirical barriers in the validation of tests for clinical use that must be overcome before the larger question of clinical utility can be tackled include the interdisciplinary linkages between the "-omic" sciences and nutritional sciences which have not hardened into a theoretical synthesis, assay methodologies and data collection techniques which are not standardized, and the clinical application of nutrigenetics.

Although none of these observations will take the wind out of the sails of either nutrigenomics or nutrigenetics, they do suggest that our original question, "Is the science of nutrigenomics and nutrigenetics sufficiently strong to warrant claims of health benefits?", could be answered only with a heavily qualified "yes" or a precautionary "no." The reasons become more apparent when we consider in detail a concrete example of nutrigenomics and nutrigenetics, the role of genetic mutations for methylene tetrahydrofolate reductase (MTHFR), folate, and the risk of cardiovascular disease. MTHFR has been studied extensively as a nutrigenetic phenomenon, and tests for genetic variants of the genes involved have been offered through private companies. MTHFR is the enzyme that is responsible for the regulation of homocysteine levels in the body. Normally, MTHFR is responsible for reducing one form of folate, methylene tetrahydrofolate, to another form, methyl tetrahydrofolate. This second form of folate donates a methyl group to homocysteine in a reaction catalyzed by the enzyme methionine synthase.

There are two known polymorphisms of MTHFR. One involves the substitution of a cytosine for an adenine base (1298A > C), and the other involves the substitution of a thymine for a cytosine base (677C > T). The 1298A > C point mutation changes the amino acid at position 429 of the MTHFR molecule from glutamine to alanine, whereas the 677C > T point mutation causes the substitution of alanine by valine at position 222 in the MTHFR molecule. Because the 1298A > C mutation leads to a structurally unstable form of MTHFR which tends to dissociate into monomers, it is not well characterized (Yamada et al. 2001). The 677C > T polymorphism is more structurally stable but is still somewhat more thermolabile, which is the cause of reduced MTHFR activity for individuals with the 677C > T

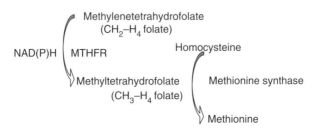

Figure 2.1. Role of MTHFR in homocysteine regulation.

polymorphism (Frosst et al. 1995). Of the approximately 10 to 15 percent of the population that has this polymorphism, all have elevated levels of homocysteine. To compensate for the decreased MTHFR enzymatic activity, increased folate intake ensures that homocysteine does not accumulate (De Bree et al. 2002).

Folate intake is known to have many implications for development and homeostasis. Supplements are widely endorsed for pregnant woman who have low blood folate levels to reduce the risk of giving birth to a child with neural tube defects, a condition that carries high mortality and morbidity rates for the child. Folic acid supplements have been shown to decrease the rate of neural tube defects in newborns (Medical Research Council Vitamin Study Research Group 1991). Elevated homocysteine levels (homocysteinuria) are associated with increased risk for Down's syndrome, neural tube defects, mental disorders, and cancer (Ames et al. 2002). Homocysteinuria is also regarded as a risk factor for cardiovascular disease (CVD) (Boushey et al. 1995; Graham et al. 1997), so a similar strategy of folate supplementation might be pursued to mitigate the risk of CVD (Ashfield-Watt et al. 2002).

Folate's wide range of impact but variable effects makes it an excellent study for nutrigenomics, particularly since SNPs have been identified, some metabolic pathways characterized, and some clinical studies have been conducted. Folate might be the best candidate for closing the loop between nutrigenomics and personalized nutrigenetics (Stover and Garza 2002). When combined with a simple genetic test for the 677C > T polymorphism, it would appear that nutrigenomics has come of age.

Some recent studies of the relationship between MTHFR polymorphisms, homocysteine levels, and CVD risk lend support to this idea, but the strength of the support is not overly convincing. The Homocysteine Studies Collaboration, for example, conducted a metaanalysis of 30 other studies, from which they concluded that homocysteine is a "modest independent predictor" of increased CVD risk (Homocysteine Studies Collaboration 2002). This result was corroborated by a European study conducted in 2003 (Meleady et al. 2003). Another metaanalysis suggested that for those who have the 677C>T polymorphism, the risk of CVD was 16 percent higher than for the normal variant (Klerk et al. 2002). Other research is, however, less conclusive about the relationship between elevated levels of homocysteine and CVD. Ma et al., for example, report that risk of myocardial infarction is not increased in a cohort study of MTHFR polymorphisms (Ma et al. 1996). A recent retrospective study of stroke patients who had their homocysteine levels reduced found that the reduction did not lead to reduced stroke or myocardial infarction, despite finding a constant association of higher than normal homocysteine levels and vascular risk (Toole et al. 2004). Another recent prospective study does not support the claim that high homocysteine levels are associated with increased CVD, but rather, that elevated levels of folate are, independent of homocysteine, responsible for reduced CVD risk (Voutilainen et al. 2004). As the authors note, the group at risk of CVD had 10 percent lower serum folate, whereas those who had higher folate levels were strongly associated with reduced risk, even after other dietary factors and CVD risk factors were accommodated. Finally, a case–control study suggested that the 677C > T polymorphism posed no risk of CVD even when folate levels are low (Verhoef et al. 1998).

A genetic test for 677C > T variant individuals is easy, inexpensive, and has been offered directly to consumers. Presumably, all of the expected 15 percent of the population with the MTHFR polymorphism would have received dietary advice to increase intake of folate. Given inconclusive reports about homocysteine as a biomarker for susceptibility to CVD, it is unclear whether increased folate intake is warranted. Perhaps in this particular nutrigenetic case the folate intervention carries no known additional risks, so a "better safe than sorry" strategy makes sense. Evidence-based medicine is not, however, attuned to this approach as either a therapeutic or preventive strategy. Accordingly, the clinical validity of this nutrigenetic intervention needs the support of large-scale prospective trials in which homocysteine is lowered by means other than folate (Refsum 2004).

2.4   ROOM FOR IMPROVEMENT

As the case study of MTHFR suggests, there is strong evidence correlating certain SNPs, and their association with particular nutrients, in downstream metabolic disturbance implicated in disease progression. Yet in the case of MTHFR, the strength of the connection between the presence of the SNP and the development of disease now or in the future is debatable. MTHFR is one of many enzymes involved in the pathway, of which not all of the pathway's functions are clearly understood. MTHFR is the exception in the pathway, since not all of the rest of the enzymes have known genetic variants, and a smaller number still have had differential activity associated with nutrients.

Despite the fact that MTHFR is often lauded as a paradigmatic example of nutrigenomics, this exemplar is associated with some uncertainties. But what does that mean? It means that on the one hand, practitioners of nutri-*genomics*, as a new field of science, have considerable amounts of spade-work to do before nutrient–gene associations and disease susceptibilities are well understood for a broad number of cases, including ApoE, for instance

(Masson et al. 2003). This observation is linked to the *completeness criticism* of nutrigenomics—the science is in its infancy and is incomplete. On the other hand, if there are more knowledge gaps than knowledge bridges in nutrigenomics, it suggests that evidence supporting claims about gene–nutrient interaction and disease progression are incomplete and could be undermined by future studies. This observation about nutri*genetics* is linked to the *soundness criticism*—the science is too young to warrant predictions about individual disease susceptibility or premature commercialization (Vineis and Christiani 2004).

The completeness and soundness criticisms are obviously linked, but in the scientific literature a thematic approach to evaluating nutrigenomics and nutrigenetics is substituted with recommendations for improving different facets of the field. These, considered below, comprise a panel of potential areas for improving the strength of the science. We will return to the issue of completeness and soundness in nutrigenomics in Section 2.5.

2.4.1 Study Design

One of the greatest challenges facing nutritional genomics arises from the relative infancy of the science. Although nutrient–gene interactions have been studied for several decades, nutrigenomics has only recently been recognized as an established discipline. This indicates that significant funding for the field as such is recent. Consequently, many of the candidate gene analyses and association studies have been retrospective, often making secondary use of existing data such as that accrued in the Framingham Heart Study. Ordovas and Mooser point out in a recent discussion of nutrigenomics, what is currently missing are the "large prospective population studies with carefully collected behavioral, clinical, biochemical, and clinical data" (Ordovas and Mooser 2004) that would ultimately generate a more robust nutrigenomic science. Without these data it becomes impossible to solve a problem that poses a serious threat to clinical applications of nutrigenetics.

Apart from the role that funding would play in relieving researchers of the need to make secondary use of existing data, prospective studies would present their own challenges. Loktionov (2003) points out that as association studies move toward candidate genes that appear to have multiple functions and several variants, studies involving simultaneous analysis of several genes will require higher levels of environmental controls—which, if these were to be conducted with volunteer subjects, would require precise dietary intake assessments and a greater number of biomarkers. The complexity of these types of studies suggest that improved consistency in reporting of epidemiological, genomic, and genetic information will become crucial (Khoury 2002).

Another dimension to the design of studies moves away from the genetics of nutrient–gene interactions. Forman et al. (2004) have suggested that the study of nutrition and cancer prevention will proceed more rapidly if new study designs are considered in which there is crossover between human and nonhuman trial subjects. Their contention is that nutritional genomics needs to become *more* interdisciplinary, to take advantage of potential synergies between animal models and human trials.

2.4.2   Epigenetics

Epigenetics, which is the study of the alteration of gene expression without changes to the genes themselves, is an intensive area of research in genomics (Jackson 2003). It will be particularly important for nutritional genomics, with mounting evidence that nutrients and other metabolites play important roles in gene expression. Although some regard the data as inconclusive (Kim 2004), there is some evidence that low folate is implicated in cancer development via DNA methylation, for example, which places nutrigenomics squarely in one of the most significant trends in cancer research (Jones 2003). Studies indicating that some nutrients have anticancer properties, for example, are now recasting correlations between diet and disease frequency in terms of epigenetic causal mechanisms (Ross 2003). The broad implication is that not only does exposure to nutrients affect the propensity for disease and longevity within a lifetime, it may, because some epigenetic phenomena accrued in a lifetime appear to be heritable, have intergenerational effects. If this is true, the implications for the nutritional sciences are profound (Mathers 2005).

2.4.3   SNPs and Haplotypes

Although nutrigenomics and nutrigenetics may be relatively new as an integrative scientific field, it has at its disposal very powerful tools arising from the Human Genome Project. In addition to microarray technology and bioinformatics (Gohil and Chakraborty 2004), nutrigenomics involves the use of single nucleotide polymorphism (SNP) analysis, gene expression profiling, proteomics (proteomics), metabolomics, bioinformatics, biocomputation, and the aid of the International HapMap Consortium (Muller and Kersten 2003). The systems biology approach is generating new biomarkers that are crucial in the measurement of nutrient–gene interactions and in identifying the contributions of individual genes to diseases that have a complex multigene basis (Simopoulos 2002; van Ommen 2004). Bearing in mind that there are more than 10 million SNPs throughout the human genome, extensive genotyping can be expensive and difficult. However, the relatively recent discovery of the haplotype structure of SNP inheritance could greatly simplify the task of defining those genetic variations associated with nutrition and metabolism. The HapMap project may provide the knowledge to get a deeper understanding of this subject, and cost-efficiencies from that knowledge could provide major medical benefits. Launched in 2003, the HapMap project is a major international effort mapping patterns of DNA sequence variation by determining the genotypes of more than a million sequence variants, their frequencies, and the degree of association between them in DNA samples from people from Africa, China, and Japan, and those of northern and western European ancestry. When the HapMap is complete, it is estimated that just 300,000 to 600,000 "tag" SNPs will be sufficient to define the most significant genetic variation. Compared to the prospect of having to type all 10 million SNPs, this represents a dramatic reduction in time and effort. Genotyping only a few carefully chosen SNPs in a chromosome region will be enough to predict the remainder of the nearby common SNPs (International HapMap Project 2006).

2.4.4   Dietary Intake Assessment

A potentially powerful source of information in nutritional genomics consists of intake reports of people participating in research studies. Unfortunately, there is an inverse relationship between the reliability of intake data and the cost and size of the study. Self-reports are notoriously unreliable, because people commit a litany of misreporting, including underreporting of intake of foods they know or perceive to be "bad" while overreporting the frequency and quantity of foods they consider to be "good." The error rates, sometimes ranging in the hundreds of percentage points, can be established in self-report trials using heavy water–labeled foods (Bingham 1991). The virtue of self-reports is that large numbers of volunteers, or minimally paid subjects, can often be recruited. The value of large data sets is therefore frequently exchanged for a precise data set. The latter is obtainable in controlled-feeding trials. These are expensive to administer to large groups, which can have a negative effect on studies aiming to establish nutrient–gene interactions for large populations or to distinguish in-group variability within populations.

2.4.5   Biomarkers

One of the stumbling blocks for nutrigenetics lies within the nutritional sciences themselves. Until very recently, nutritional intervention studies have not integrated genomic, proteomic, or metabolomic analyses (Daniel 2002; Kim et al. 2004). The effect of interventions can be observed and measured in a clinical context, but without the aid of the powerful biomarkers becoming available through systems biology research (van Ommen and Stierum 2002). Now that nutrition research is incorporating genomic, proteomic, and metabolomic principles and techniques, the question is whether these techniques will safely and effectively lead to more precise dietary advice (Mensink and Plat 2002; Muller and Kersten 2003).

Lenore Arab and others have argued that much of the data and techniques needed to calibrate the generalized knowledge about nutrient–gene interactions to individuals are still in the future (Arab 2004). In part, this is because of a phenomenon common to all systems biology: Many of the biomarkers appropriate for "personalized medicine" have not been developed or appropriately coordinated with drug development or nutrition intervention studies (Meyer and Ginsburg 2002; Milner 2002; Gibney et al. 2005). Even when they are, it is doubtful that physicians and nutritionists who lack the requisite backgrounds in genetics and molecular nutrition will be able to make immediate and widespread use of nutrigenetics and offer personalized services.

2.4.6 Susceptibility and Predictions

Some have raised concerns about the current usefulness of nutrigenetic tests as tools for health promotion and disease prevention because the science is incomplete, so sound predictions cannot be made (Khoury 2003). With nutrigenetics, the focus is on establishing a person's genetic profile in the hopes that the predictive value of testing for single "key" mutations will help to determine individual nutrient requirements as well as susceptibility to particular nutrition-related diseases. According to Nancy Fogg-Johnson and Jim Kaput, "*if* genetic tests were available for the variant gene and if that variant was shown to be the only cause of a disease process, a physician or nutritional expert could recommend increasing or decreasing intake of a specific vitamin or food" (Fogg-Johnson and Kaput 2003). Given the drive to associate SNPs with diseases, it is tempting to conclude that the presence of an SNP or set of SNPs indicates disease susceptibility. But where large disease-association studies have been conducted for cardiovascular disease, it appears that some SNPs may be associated with disease but only for a subset of the population that has the SNP (Ordovas 2002). Thus, even though the SNP can *in general* be associated with a disease, further disease-association assays are necessary to confirm that *this particular person* is a member of the subset at risk.

2.4.7 Analytical and Clinical Validity

Nutrigenetics ultimately depends on high-quality genetic tests that will reliably indicate which nutritional interventions will have positive health outcomes. The consistency and reliability of tests using common commercial microarray platforms is a known issue (Tan et al. 2003). The U.S. Centers for Disease Control convened a workshop in 2001 to determine the quality of evidence required when assessing the appropriate use of new genetic tests. The participants of the workshop concluded that to adequately meet the information needs of clinicians, policymakers, consumers, and patients, genetic tests should be subject to a standardized system of clinical evaluation. As documented by Burke et al. (2002), among the criteria the group identified as necessary for an objective assessment of a given test's properties are its analytic validity, its clinical validity, and its clinical utility:

> *Analytic validity* refers to the accuracy with which a particular genetic characteristic—probably a SNP—can be identified in a given laboratory test. Analytic validity may be expressed in terms of a test's sensitivity, specificity, reproducibility and predictive value, but because more than one test can often be used to find genetic variants, comparison studies are needed before a professional consensus on methods for nutrigenetic testing will emerge.
>
> *Clinical validity* describes the accuracy with which a test predicts a particular clinical outcome—improved health or disease prevention—based on a link between the diseases and the genetic mutations being analyzed. Even when a sufficient number of studies have been completed, nutrigenetic tests may remain imprecise because of complex gene–gene and gene–environment interactions. Uncertainty about clinical validity is an important consideration because people need to know if the information is sound enough to justify such actions as changes in diet.

Together, these criteria should ensure that as an evidence-based practice, nutrigenetics "should increasingly rely on scientific data on analytic performance of [genetic] information, its validity in predicting health outcomes, and its utility in improving health and preventing disease beyond approaches that do not use genetic information" (Khoury 2003). One major critical review of the field conducted by the Nuffield Trust in conjunction with the Public Health Genetics Unit at Cambridge University was skeptical about the clinical validity of nutrigenetics, concluding that "there is no evidence at present to support clinical applications involving individualized dietary advice based on gene testing" (Nuffield Trust 2005).

2.4.8 Clinical Utility

Related to analytical and clinical validity is the concept of *clinical utility*, which can be defined as follows: "*Clinical utility* refers to the likelihood that the test will lead to an improvement in health outcomes. Since measurement of this utility requires the evaluation of outcomes associated with clinical interventions, the quality of upstream nutritional interventions studies plays an important role" (Burke et al. 2002). Even if it is widely acknowledged that lifestyle change and dietary changes could have profound effects on disease susceptibility, these benefits will be lost as long as people do not alter their behavior. Nutrigenetics would have clinical utility if, in addition to uncovering risks greater than those associated with poor lifestyle and diet, the *genetic* aspect of the information would motivate behavioral change. One suggestion has been to develop genomic profiles for patients in which multiple gene variants implicated in pathogenesis would be detected and presented *en masse* to the patient. Haga et al. suggest that "the profiles are proposed as a means to identify individual risk, for the purpose of tailoring specific risk-reducing actions, typically involving vitamins, environmental exposures, diet or other lifestyle changes that are expected to prevent disease" (Haga et al. 2003). Such profiling tools would be clinically useful if they can effectively answer the following concerns:

1. Will people with a positive test result make the suggested lifestyle changes to improve their health?
2. Will people with a negative test result have less motivation to pursue healthy lifestyles?
3. Are lifestyle recommendations dependent on genotype information?

Haga et al. conclude that because the analytical and clinical validity of nutrigenetic tests are for the most part presently inconclusive, they do not provide grounds for thinking that people will find much utility in them either. This skepticism could be overturned if the front line of nutritional advice and counseling, dieticians, are trained appropriately in nutritional genomics (DeBusk et al. 2005).

2.5 SCIENCE AND TECHNOLOGY ASSESSMENT

In the preceding section we discussed a number of facets of nutritional genomics in which scientists working in the field have identified some shortcomings. It comes as no surprise that a relatively new and rapidly growing

---

Room for Improvement

Limitations in study design
Need to incoporate epigenetics
Improvements to SNP identification and haplotyping
Problems associated with dietary intake assessment
Need for better biomarkers
Stronger claims about susceptibility and predictions
Need to demonstrate analytical and clinical validity
Need to demonstrate clinical utility

---

field of science and technology should be occasioned with review articles and editorials in scientific journals reflecting on the field's direction. One might say that *about* nutritional genomics, at the level of general theoretical commitments, there is consensus that it is a promising field. *Within* the field, there is disagreement about the quality of experimental research, the data produced, and the reliability of predictions that can be made. The question, then, is: What can one say about the state of nutritional genomics if it is prone to completeness and soundness criticisms from within the ranks?

Formal science and technology assessments are frequently conducted in order to respond to questions about the strength of a science, particularly when the science is being applied or when the intention is to find practical applications of the science in the near future. Science and technology assessments evaluate the quality of science and consider whether its use will generate new social risks that need to be managed, possibly through regulation. By establishing a connection between the strength of the science (and its technical applications) and the potential for social risk, a useful framing assumption is being introduced that should be drawn out and clarified. The assumption is that "scientific research and technological developments may help to solve social problems, but the activity of research and development also generates social problems that cannot be separated from the R&D process itself" (Ravetz 1970).

How these social problems are engaged is a jurisdiction-specific matter. For example, assessments of technology such as the integrative technology assessments conducted in the European Union are interdisciplinary analyses that gather different disciplinary perspectives into a project-relevant framework (Decker 2004; Decker and Ladikas 2004). These technology assessments capitalize on a number of different social interaction methodologies, such as consensus workshops or the Delphi method, in order to determine to what extent there is agreement on the status of the science and technology in question. One common and effective approach now being used brings the science peer group together with another group tasked with determining and weighing social issues based on their limited, not-expert knowledge of the science and technology in question. Sometimes described as *direct confrontation methodology*, this technology assessment method is intended to generate new perspectives that are not restricted to specific disciplines. Rather, the intention is that the new perspectives will emerge from the dialogue of the participants in which the sources of scientific and social-scientific expertise are challenged. This methodology is committed to putting scientific and social considerations on a par in a matrix, in order to understand their interaction and the implications of science and technology innovation. A desirable feature that results from this process is that normative and nonscientific considerations can shape the technology assessment as easily as the strictly scientific factors.

Here it is worthwhile to distinguish between endogenous and exogenous sources of social risk. *Endogenous risks* come from two main sources. The first source is the inherent complexity of nutrigenomics science and the various subfields on which it is based. One way to think about this is to consider that if the natural processes studied in nutrigenomics are irreducibly complex, phenomena in nutrigenomics will appear to exhibit stochastic behavior that would make it a science suitable for statistical modeling. Modeling would work only to a point, however, after which at it would be clear that predictions would not be supportable. This would be due to, for example, the effects of random genetic effects: the uncertainty generated in clinical trials in which there is unpredictable assimilation of information by research subjects, coupled with a low likelihood that people will act consistently on new information. The second source of uncertainty arises from a lack of knowledge about nutrigenomics science, which can be divided into instances in which the information held is considered unreliable because it is inexact or at the limits of statistical significance, or because there are structural uncertainties caused by ignorance or conflicting interpretations of the available data. Both of these situations are characterized in the primary nutrigenomics literature.

In contrast to the endogenous risks, the *exogenous risks* are strictly social risks. These pertain to the ethical, economic, and legal effects, intended or not, which are posed by the development and application of nutrigenomics. There are a number of cognate social problems that have arisen in related cases of science and technology development which may arise in nutrigenomics. Other problems unique to nutrigenomics are also predicted. Recommendations of the type that we make in Chapter 7 can be made about the most appropriate and feasible normative, legal, and economic solutions for each of these sources of potential social disruption and risk (Castle et al. 2006). For example, nutrigenomics requires careful management of biological samples and personal information in research, clinical, and direct-to-consumer contexts. Specific guidelines, such as those being developed by the European Nutrigenomics Organisation (NuGO), may be necessary for research. Specific clinical and commercial counterparts may also have to be developed. Moreover, the provision of nutrigenomic services will vary across jurisdictions, making it impossible to recommend a one-size-fits-all regulatory strategy for nutrigenomics. In general, the single greatest challenge will be to understand what health claims are considered to be substantiated by the existing science. It is also anticipated that like other biotechnologies, there will be differential access to nutrigenomics, and this could represent an aggravation of existing disparity in access to biotechnologies for health.

To address disagreement about how well substantiated nutritional genomics is and how to ensure that only the best of the science is available to the public, it may be necessary to establish scientific guidelines for the provision of nutrigenetic tests and services. This may involve evidence-based clinical practice guidelines of the sort described above that would address the use of particular genetic tests in particular health care settings. These guidelines would be jurisdiction-based and may require new governance mechanisms for a complex field such as nutritional genomics. Some measure of consistency could be achieved if the guidelines take into account such common factors as the nature of the test involved, the quality of research on which the test is based, the value of the test for the population being served, the cost of the test, the test's acceptability in the setting in question, the test's priority relative to other health care services, and the health outcomes generated by the test compared with outcomes that could be generated by non-genetic means. Since most nutrigenetic testing is currently offered through health practitioners, these guidelines would have a wide reach. The guidelines would, however, have less, possibly no impact if nutrigenetic testing is offered outside a clinical context. Whichever setting is offering nutrigenetic testing, evidence analysis using a formally acceptable methodology for systematic reviews should be made available for adequate characterization of the state of the science for each test proposed.

2.6 CONCLUSION

The science of nutrigenomics and nutrigenetics is evolving rapidly and holds great promise for the detection of susceptibilities to disease and for health improvements that link genes to nutrition. Its promoters speak of benefits, and a number of companies are marketing services. Nutritional genomics is, however, an emerging science. Analyzing the risks and benefits of these interventions requires systematic observation in the form of well-designed controlled trials, cohort studies, or case–control studies (Burke et al. 2002). So far, the scarcity of nutritional intervention studies and the lack of information regarding the analytic and clinical validity of nutrigenetic tests means that it is not yet possible to proclaim the overall utility of such tests (Melzer and Zimmern 2002).

As a result, it is fair to recommend that formal science and technology assessments ought to be conducted, and that these should be "rolling" (i.e., conducted periodically) to keep abreast of developments in the science and technology. This continuous assessment of the science of nutrigenomics and nutrigenetics and their applications will be essential to the development of appropriate regulations and ethical guidelines which will ensure that the maximum benefits of nutritional genomics are made public and timely, with assurances that foreseeable risk has been mitigated.

REFERENCES

Ames, B. N., I. Elson-Schwab, and E. A. Silver. 2002. High-dose vitamin therapy stimulates variant enzymes with decreased coenzyme binding affinity [increased $K(m)$]: relevance to genetic disease and polymorphisms. *American Journal of Clinical Nutrition* 75(4):616–658.

Arab, L. 2004. Individualized nutritional recommendations: Do we have the measurements needed to assess risk and make dietary recommendations? *Proceedings of the Nutrition Society* 63:167–172.

Ashfield–Watt, P. A. L., C. H. Pullin, J. M. Whiting, Z. E. Clark, S. J. Moat, R. G. Newcombe, M. L. Burr, M. J. Lewis, H. J. Powers, and I. F. W. McDowell. 2002. Methylenetetrahydrofolate reductase 677C→T genotype modulates homocysteine responses to a folate-rich diet or a low-dose folic acid supplement: a randomized controlled trial. *American Journal of Clinical Nutrition* 76:180–186.

Bingham, S. A. 1991. Limitations of the various methods for collecting dietary intake data. *Annals of Nutrition and Metabolism* 35:117–127.

Boushey, C. J., S. A. Beresford, G. S. Omenn, and A. G. Motulsky. 1995. A quantitative assessment of plasma homocysteine as a risk factor for vascular disease: probable benefits of increasing folic acid intakes. *Journal of the American Medical Association* 274(13): 1049–1057.

Brennan, R. O. *Nutrigenetics: New Concepts for Relieving Hypoglycemia.* New York: M. Evans and Company.

Bubela, T. M., and T. A. Caulfield. 2004. Do the print media "hype" genetic research? A comparison of newspaper stories and peer-reviewed research papers. *Canadian Medical Association Journal* 170(9):1399–1407.

Burke, W., D. Atkins, M. Gwinn, A. Guttmacher, J. Haddow, J. Lau, G. Palomaki, N. Press, C. S. Richards, L. Wideroff, and G. L. Wiesner. 2002. Genetic test evaluation: information needs of clinicians, policy makers, and the public. *American Journal of Epidemiology* 156(4):311–318.

Caspi, A., J. McClay, T. E. Moffitt, J. Mill, J. Martin, I. W. Craig, A. Taylor, and R. Poulton. 2002. Role of genotype in the cycle of violence in maltreated children. *Science* 297(5582):851–854.

Cassman, M. 2005. Barriers to progress in systems biology. *Nature* 438:1079.

Castle, D., C. Cline, A. S. Daar, C. Tsamis, and P. A. Singer. 2006. Nutrients and norms: ethical issues in nutritional genomics. In *Nutrigenomics: Concepts and Technologies*, edited by J. Kaput and R. L. Rodriguez. Hoboken, NJ: Wiley.

Cooper, R. S., and B. M. Psaty. 2003. Genomics and medicine: distraction, incremental progress, or the dawn of a new age? *Annals of Internal Medicine* 138(7):576–580.

Daniel, H. 2002. Genomics and proteomics: importance for the future of nutrition research. *British Journal of Nutrition* 87(Suppl. 2):S305–S311.

Davis, C. D., and J. Milner. 2004. Frontiers in nutrigenomics, proteomics, metabolomics and cancer prevention. *Mutation Research* 551(1–2):51–64.

DeBree, A., W. M. Verschuren, L. A. Kluijtmans, and H. J. Blom. 2002. Homocysteine determinants and the evience to what extent homocystenine determines the risk of coronary heart disease. *Pharmocological Review* 54:599–618.

DeBusk, R. M., C. P. Fogarty, J. M. Ordovas, and K. S. Kornman. 2005. Nutritional genomics in practice: Where do we begin? *Journal of the American Dietetic Association* 105:589–598.

Decker, M. 2004. *Interdisciplinarity in Technology Assessment: Implementation and Its Chances and Limits*. New York: Springer-Verlag.

Decker, M., and M. Ladikas. 2004. *Bridges Between Science, Society and Policy*. New York: Springer-Verlag.

Elliott, R., and T. J. Ong. 2002. Nutritional genomics. *British Medical Journal* 324(7351):1438–1442.

Evans, W. E., and J. A. Johnson. 2001. Pharmacogenomics: the inherited basis for interindividual differences in drug response. *Annual Review of Genomics and Human Genetics* 2:9–39.

Evans, W. E., and H. L. McLeod. 2003. Pharmacogenomics: drug disposition, drug targets, and side effects. *New England Journal of Medicine* 348(6):538–549.

Fairweather-Tait, S. J. 2003. Human nutrition and food research: opportunities and challenges in the post-genomic era. *Philosophical Transactions of the Royal Society of London, B: Biological Science* 358(1438):1709–1727.

Fogg-Johnson, N., and J. Kaput. 2003. Nutrigenomics: an emerging scientific discipline. *Food Technology* 57:60–67.

Forman, M. R., S. D. Hursting, A. A. Umar, and J. C. Barrett. 2004. Nutrition and cancer prevention: a multidisciplinary perspective on human trials. *Annual Review of Nutrition* 24.

Frosst, P., H. J. Blom, R. Milos, P. Goyette, C. A. Sheppard, R. G. Matthews, G. J. Boers, M. den Heijer, L. A. Kluijtmans, L. P. van den Heuvel, et al. 1995. A candidate genetic risk factor for vascular disease: a common mutation in methylenetetrahydrofolate reductase. *Nature: Genetics* 10(1):111–113.

Gibney, M. J., M. Walsh, L. Brennan, H. M. Roche, B. German, and B. van Ommen. 2005. Metabolomics in human nutrition: opportunities and challenges. *American Journal of Clinical Nutrition* 82:497–503.

Gillies, P. J. 2003. Nutrigenomics: the Rubicon of molecular nutrition. *Journal of the American Dietetic Association* 103(12, Suppl. 2):S50–S55.

Gohil, K., and A. A. Chakraborty. 2004. Applications of microarray and bioinformatics tools to dissect molecular responses of the central nervous system to antioxidant micronutrients. *Nutrition* 20(1):50–55.

Graham, I. M., L. E. Daly, H. M. Refsum, K. Robinson, L. E. Brattstrom, P. M. Ueland, R. J. Palma-Reis, G. H. Boers, R. G. Sheahan, B. Israelsson, et al. 1997. Plasma homocysteine as a risk factor for vascular disease: the European Concerted Action Project. *Journal of the American Medical Association* 277(22):1775–1781.

Haga, S. B., M. J. Khoury, and W. Burke. 2003. Genomic profiling to promote a healthy lifestyle: not ready for prime time. *Nature: Genetics* 34(4):347–350.

Holtzman, N. A. 1998. Recombinant DNA technology, genetic tests and public policy. *American Journal of Human Genetics* 48:624–632.

Homocysteine Studies Collaboration. 2002. Homocysteine and risk of ischemic heart disease and stroke: a meta-analysis. *Journal of the American Medical Association* 288:2015–2022.

Hood, L. 2003. Systems biology: integrating technology, biology, and computation. *Mechanisms of Ageing and Development* 124(1):9–16.

International HapMap Project. 2006. Retrieved June 20, 2006 from www.hapmap.org.

Jackson, D. A. 2003. The principles of nuclear structure. *Chromosome Research* 11:387–401.

Jimenez-Sanchez, G., B. Barton Childs, and D. David Valle. 2001. Human disease genes. *Nature* 409:853–855.

Jones, P. 2003. Epigenetics in carcinogenesis and cancer prevention. *Annals of the New York Academy of Sciences* 13:511–519.

Kaput, J. 2004. Diet–disease gene interactions. *Nutrition* 20(1):26–31.

Kaput, J., and R. L. Rodriguez. 2004. Nutritional genomics: the next frontier in the postgenomic era. *Physiological Genomics* 16(2):166–177.

Khoury, M. J. 2002. Commentary: Epidemiology and the continuum from genetic research to genetic testing. *American Journal of Epidemiology* 156:297–299.

———. 2003. Genetics and genomics in practice: the continuum from genetic disease to genetic information in health and disease. *Genetics and Medicine* 5(4):261–268.

Kim, H., G. P. Page, and S. Barnes. 2004. Proteomics and mass spectrometry in nutrition research. *Nutrition* 20(1):155–165.

Kim, Y.-I. 2004. Folate and DNA methylation: a mechanistic link between folate deficiency and colorectal cancer? *Cancer Epidemiology, Biomarkers and Prevention* 13:511–519.

Klerk, M., P. Verhoef, R. Clarke, H. J. Blom, F. J. Kok, and E. G. Schouten. 2002. MTHFR 677C → T polymorphism and risk of coronary heart disease: a meta-analysis. *Journal of the American Medical Association* 288(16):2023–2031.

Koch, R. K. 1999. Issues in newborn screening for phenylketonuria. *American Family Physician* 60(5):1462–1466.

Kornman, K. S., P. M. Martha, and G. W. Duff. 2004. Genetic variations and inflammation: a practical nutrigenomics opportunity. *Nutrition* 20(1):44–49.

Lander, E. S., L. M. Linton, B. Birren, C. Nusbaum, M. C. Zody, J. Baldwin, K. Devon, K. Dewar, M. Doyle, W. FitzHugh, et al. 2001. Initial sequencing and analysis of the human genome. *Nature* 409(6822):860–921.

Loktionov, A. 2003. Common gene polymorphisms and nutrition: emerging links with pathogenesis of multifactorial chronic diseases. *Journal of Nutritional Biochemistry* 2003:426–451.

Ma, J., M. J. Stampfer, C. H. Hennekens, P. Frosst, J. Selhub, J. Horsford, M. R. Malinow, W. C. Willett, and R. Rozen. 1996. Methylenetetrahydrofolate reductase polymorphism, plasma folate, homocysteine, and risk of myocardial infarction in US physicians. *Circulation* 94(10):2410–2416.

Masson, L. F., G. McNeill, and A. Avenell. 2003. Genetic variation and the lipid response to dietary intervention: a systematic review. *American Journal of Clinical Nutrition* 77: 1098–1111.

Mathers, J. C. 2005. Nutrition and epigenetics: how the genome learns from experience. *British Nutrition Foundation Nutrition Bulletin* 30:6–12.

Mathew, C. 2001. Science, medicine, and the future: postgenomic technologies—hunting the genes for common disorders. *British Medical Journal* 322(7293):1031–1034.

Medical Research Council Vitamin Study Research Group. 1991. Prevention of neural tube defects: results of the Medical Research Council Vitamin Study. *Lancet* 338(8760): 131–137.

Meleady, R., P. M. Ueland, H. Blom, A. S. Whitehead, H. Refsum, L. E. Daly, S. E. Vollset, C. Donohue, B. Giesendorf, I. M. Graham, et al. 2003. Thermolabile methylenetetrahydrofolate reductase, homocysteine, and cardiovascular disease risk: the European Concerted Action Project. *American Family Physician* 77(1):63–70.

Melzer, D., and R. Zimmern. 2002. Genetics and medicalisation. *British Medical Journal* 324(7342):863–864.

Mensink, R. P., and J. Plat. 2002. Post-genomic opportunities for understanding nutrition: the nutritionist's perspective. *Proceedings of the Nutrition Society* 61(4):401–404.

Meyer, J. M., and G. S. Ginsburg. 2002. The path to personalized medicine. *Current Opinion in Chemistry and Biology* 6(4):434–438.

Middleton, F. A., E. J. Ramos, Y. Xu, H. Diab, X. Zhao, U. N. Das, and M. Meguid. 2004. Application of genomic technologies: DNA microarrays and metabolic profiling of obesity in the hypothalamus and in subcutaneous fat. *Nutrition* 20(1):14–25.

Milner, J. 2002. Functional foods and health: a US perspective. *British Journal of Nutrition* 88(S2):S151–S158.

Muller, M., and S. Kersten. 2003. Nutrigenomics: goals and strategies. *Nature Reviews: Genetics* 4:315–322.

Nuffield Trust. 2005. *Nutrigenomics: Report of a Workshop Hosted by the Nuffield Trust and Organised by the Public Health Genetics Unit.* London: NT.

Ordovas, J. M. 2002. Gene–diet interaction and plasma lipid responses to dietary intervention. *Biochemistry Society Transactions* 30(2):68–73.

———. 2003. Cardiovascular disease genetics: a long and winding road. *Current Opinion in Lipidology* 14(1):47–54.

Ordovas, J., and D. Corella. 2004. Hutritional genomics. *Annual Review of Genomics and Human Genetics* 5:71–118.

Ordovas, J. M., and V. Mooser. 2004. Nutrigenomics and nutrigenetics. *Current Opinion in Lipidology* 15(2):101–108.

Peregrin, T. 2001. The new frontier of nutrition science: nutrigenomics. *Journal of the American Dietetic Association* 101(11):1306.

Ravetz, J. R. 1970. *Scientific Knowledge and Its Social Problems.* Oxford: Clarendon Press.

Refsum, H. 2004. Is folic acid the answer? *American Journal of Clinical Nutrition* 80(2): 241–242.

Ridley, M. 2003. Nature via nurture: *Genes, Experience, and What Makes us Human*. New York: HarperCollins.

Roche, H. M., C. Phillips, and M. J. Gibney. 2005. The metabolic syndrome: the crossroads of diet and genetics. *Proceedings of the Nutrition Society* 64:371–377.

Ross, S. A. 2003. Diet and DNA methylation interactions in cancer prevention. *Annals of the New York Academy of Sciences* 983:197–207.

Schmidt, C. W. 2002. Toxicogenomics: an emerging discipline. *Environmental Health Perspectives* 110(12):A750–A755.

Simopoulos, A. P. 2002. Genetic variation and dietary response: nutrigenetics/nutrigenomics. *Asia Pacific Journal of Clinical Nutrition* 11(S6).

Stover, P. J., and C. Garza. 2002. Bringing individuality to public health recommendations. *Journal of Nutrition* 132(8, Suppl.):2476S–2480S.

Tan, P. K., T. J. Downey, E. L. Spitznagel, P. Pin Xu, D. Fu, D. S. Dimitrov, R. A. Lempicki, B. M. Raaka, and M. C. Cam. 2003. Evaluation of gene expression measurements from commercial microarray platforms. *Nucleic Acids Research* 31:5676–5684.

Toole, J. F., M. R. Malinow, L. E. Chambless, J. D. Spence, L. C. Pettigrew, V. J. Howard, E. G. Sides, C. H. Wang, and M. Stampfer. 2004. Lowering homocysteine in patients with ischemic stroke to prevent recurrent stroke, myocardial infarction, and death: the Vitamin Intervention for Stroke Prevention (VISP) randomized controlled trial. *Journal of the American Medical Association* 291(5):565–575.

van Ommen, B. 2004. Nutrigenomics: exploiting systems biology in the nutrition and health arenas. *Nutrition* 20(1):4–8.

van Ommen, B., and R. Stierum. 2002. Nutrigenomics: exploiting systems biology in the nutrition and health arena. *Current Opinion in Biotechnology* 13:517–521.

Venter, J. C., M. D. Adams, E. W. Myers, P. W. Li, R. J. Mural, G. G. Sutton, H. O. Smith, M. Yandell, C. A. Evans, R. A. Holt, et al. 2001. The sequence of the human genome. *Science* 291(5507):1304–1351.

Verhoef, P., E. B. Rimm, D. J. Hunter, J. Chen, W. C. Willett, K. Kelsey, and M. J. Stampfer. 1998. A common mutation in the methylenetetrahydrofolate reductase gene and risk of coronary heart disease: results among U.S. men. *Journal of the American College of Cardiologists* 32(2):353–359.

Vineis, P., and D. C. Christiani. 2004. Genetic testing for sale. *Epidemiology* 15:3–5.

Voutilainen, S., J. K. Virtanen, T. H. Rissanen, G. Alfthan, J. Laukkanen, K. Nyyssonen, J. Mursu, V. P. Valkonen, T. P. Tuomainen, G. A. Kaplan, and J. T. Salonen. 2004. Serum folate and homocysteine and the incidence of acute coronary events: the Kuopio Ischaemic Heart Disease Risk Factor Study. *American Journal of Clinical Nutrition* 80(2):317–323.

Wood, P. A. 2004. Genetically modified mouse models for disorders of fatty acid metabolism: pursuing the nutrigenomics of insulin resistance and type 2 diabetes. *Nutrition* 20(1): 121–126.

Wright, A. F., and N. D. Hastie. 2001. Complex genetic diseases: controversy over the Croesus code. *Genome Biology* 2:1–8.

Yach, D., D. Stuckler, and K. D. Brownell. 2006. Epidemiologic and economic consequences of global epidemics of obesity and diabetes. *Nature: Medicine* 12:62–66.

Yamada, K., Z. Chen, R. Rozen, and R. G. Matthews. 2001. Effects of common polymorphisms on the properties of recombinant human methylenetetrahydrofolate reductase. *Proceedings of the National Academy of Sciences USA* 98(26):14853–14858.

3

THE ETHICS OF NUTRIGENOMIC TESTS AND INFORMATION

3.1 INTRODUCTION

In this chapter we consider the ethical and legal issues associated with the management of personal genetic information. We are orienting this discussion toward situations in which a person has taken a genetic test and received nutritional advice, irrespective of how the person has accessed the nutrigenomic service. Alternative nutrigenomic service delivery pathways are considered in Chapter 4. Here, the focus is on the ethical and legal issues that arise once this information has been generated. It is important to note that our analysis generally takes the viewpoint of the patient and the consumer: to consider the potential for them or their families to be harmed by mismanagement of information they have solicited. This approach is consistent with the orientation of most literature on genetic information management because it takes individual autonomy, and threats to it, as the central normative and legal principle. We consider other normative principles that are at play, apart from autonomy. Similarly, we also discuss the important issues of genetic information management in the context of nutrigenomics research.

Science, Society, and the Supermarket: The Opportunities and Challenges of Nutrigenomics,
By David Castle, Cheryl Cline, Abdallah S. Daar, Charoula Tsamis, and Peter A. Singer
Copyright © 2007 John Wiley & Sons, Inc.

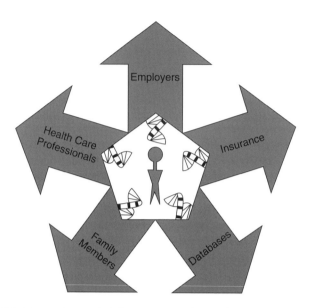

Figure 3.1. Personal genetic information can be used in a variety of contexts.

The management of personal genetic information has been a major source of ethical and legal debate over the past decade. This is true not only for clinical and private access to genetic testing, but also for genetics and genomics research. Historically, genetic testing in clinical contexts has been used in the diagnosis of serious conditions. The benefit of the genetic test is that even if there is no associated therapy, the tests themselves have had some predictive value. Clinical diagnostic tests are therefore powerful aids in providing an independent diagnosis, or supporting a diagnosis based on other considerations. Predictive genetic tests have the potential to generate definitive or at least highly reliable prognoses. Because the information from these tests can be about sensitive matters of life and death, appropriately strong controls on the information have been sought in clinical and research settings.

The tests used in nutrigenomics, by contrast, are focused on determining disease susceptibilities for low-impact genes that are not closely correlated with particular, serious conditions. That is, the low penetrance of the genetic variation that was tested for means that it neither leads, independently and decisively, to disease, nor is there acute susceptibility to disease associated with the gene variant. Because the genetic variations tested for in nutrigenomics generally do not have high predictive value but are implicated in complex pathways that can lead to disease over time, the tests are described as *susceptibility tests*. Their low predictive value and long-term impact

Genetic testing involves examining a sample of blood or other body fluid or tissue for biochemical, chromosomal, or genetic markers that indicate the presence or absence of genetic disease. Genetic tests can be used to look for possible predisposition to disease as well as to confirm a suspected mutation in an individual or family.

Genetic screening refers to programs designed to identify persons within a subpopulation whose genotypes suggest that they or their offspring are at a higher risk for a genetic disease or condition. Genetic screening programs use genetic tests (among other techniques) to identify rates of genetic diseases.

Genetic counseling provides individuals or families with education, information, and support in relation to genetic conditions and helps them make informed decisions. Genetic counseling is traditionally nondirective; that is, it provides sufficient information to allow individuals or families to determine the best course of action for them without making testing recommendations. Counseling can occur before or after a test has been administered.

means that a wellness focus replaces immediate therapy if there is a known dietary intervention that may reduce susceptibility. Although the efficacy of the dietary changes are not well documented, many of the interventions recommended are consistent with widely accepted general nutritional guidelines. In this respect, the information currently generated by nutrigenomic tests tends to be nonstigmatizing, and interventions are often relatively uncontroversial.

Given this understanding of nutrigenomic tests and the direction that current nutrigenomic research appears to be moving, we will treat nutrigenomic testing as a form of discretionary testing of a nonsensitive nature relative to predictive genetic tests, and we will regard them as having mild to moderate informational impact. Our stance, then, is that the ethical issues identified as being of concern for genetic testing in nutrigenomics, in particular the potential harm against which consumers must still be safeguarded, are often less acute in the case of nutrigenomics than in clinical medicine.

This position does not lead to the endorsement of a cavalier attitude about genetic information arising in nutrigenomics; just the opposite is the case. In its current state, nutrigenomics trades primarily in wellness and susceptibility tests. Yet nutrigenomics is a rapidly developing field, and the nature of the tests can be expected to evolve. Whether our analysis of the potential harm arising from nutrigenomic information mismanagement holds for very long depends largely on how the science unfolds. We presume that some claims made here will not hold in the future, particularly if research in nutrigenomics has greater success in providing evidence for nutrient–gene

In some cases, the results of a genetic test can be devastating. The BRCA1 and BRCA2 genetic anomalies are associated with an increased risk of breast cancer. Although a positive test does not mean a certainty of cancer, it does indicate a higher risk. Women carrying these genes are advised to undergo constant monitoring and frequent testing, and may be advised to have a double mastectomy to reduce the risk of cancer developing. In some countries, they could lose access to health insurance, and it could affect their personal relationships. They also have to decide whether to share the news of the risk with sisters and daughters, who may also be at risk.

interactions and interventions to mitigate or prevent the onset of disease. As nutrient–gene associations become better characterized and the empirical evidence for their use grows stronger, it may be that nutrigenomics will migrate from a science of susceptibility to a science of predictibility. In that case, the potential for harmful impact of information management will grow in proportion to the strengthening of the science and the consensus around its use.

In this chapter we consider issues that are raised by nutrigenomic testing. These include consent, confidentiality, familial consequences, the testing of children, and nonmedical uses of nutrigenomic information by interested third parties such as employers and insurers. As we discuss, one of the greatest challenges facing nutrigenomic testing is the issue of informed consent. The complexity and uncertainty of information in tests for disease susceptibility, along with the lack of education in this field for many health professionals, makes it difficult to resolve this issue quickly.

3.2 ETHICAL PRINCIPLES

In this chapter we identify the ethical principles that are central to the governance of nutrigenomics information and which will guide our subsequent discussion of particular ethical issues. The centrality of ethical principles in discussions about genetic testing is widely recognized in the bioethics literature, international conventions, domestic legislation, and health care practice. The principles we consider here are consent, privacy, confidentiality, genetic nondiscrimination, and genetic solidarity, the last being a relatively recent entrant into the discussion. Two international statements regarding ethical principles particularly relevant to the ethics of genetic information management are UNESCO's *Universal Declaration on the Human Genome*

Established in 1999 by the United Kingdom, the Human Genetics Commission (HGC) is an independent advisory body to monitor the impact of human genetics on people and health care and to advise on social and ethical issues. The work of the Advisory Committee on Genetic Testing, the Advisory Group on Scientific Advances in Genetics, and the Human Genetics Advisory Commission has been absorbed into the HGC. The commission (www.hgc.gov.uk) is to provide the government with advice on a range of genetics issues. A 2003 HGC report, *Genes Direct: Ensuring the Effective Oversight of Genetic Tests Supplied Directly to the Public*, deals with genetic tests available directly to the consumer. It examines regulation of these tests, and ethical and other issues surrounding their sale.

and Human Rights (1997) and the Council of Europe's convention on human rights and biomedicine (1997). We also draw on the 2002 report of the UK Human Genetics Commission on genetic information in our selection and characterization of core principles (Human Genetics Commission 2002). The principles chosen are aimed primarily at ensuring appropriate safeguards for the individual, although some weight is given to familial and societal interests.

CONSENT. Respect for persons requires that we acknowledge the value, dignity, and moral rights of others in all decision-making that affects their welfare. This means, in part, respecting a person's right to autonomy: the person's ability to choose how to live. This gives rise to a principle of consent, which draws from the person's goals, preferences, and values, and which may even extend to a right not to know the results of a test or a specific right to have genetic ignorance (Hayry and Takala 2001). In addition to promoting rational decision-making, consent protects people from coercion and manipulation (Goldworth 1999). Consent governs both the physical procedures of medicine, such as the administration of a genetic test, and the handling of the resulting information. Barring special considerations, genetic samples and private genetic information should not be obtained, held, or discussed with others without the subject's free and informed consent.

PRIVACY. For our purposes we are interested in four types of privacy rights: the right against unwanted physical intrusion such as mandatory testing (Greely 1998); the right to limit or withhold information from third parties, including the results of genetic tests (Anderlik and Rothstein 2001); the right of noninterference from third parties in making personal choices such as whether to undergo genetic testing (Anderlik and Rothstein 2001); and the right to insist that information conveyed to another party for a particular purpose not be disclosed to others (Greely 1998). This is normally referred to as a right to confidentiality and is discussed further below. Violations of privacy can result in any number of harms, including discrimination and stigmatization, lack of control over personal information, and the disruption of conditions for social life (Rachels 1975).

CONFIDENTIALITY. Confidentiality involves protecting a person's right to privacy in the context of a professional relationship, such as that between physician and patient. The physician has a professional obligation not to disclose privileged information about any patient to outside parties without the consent of the patient (Hayry and Takala 2001). To the extent that it is considered medical information, genetic information should be accorded at least a similar level of confidentiality protection.

GENETIC NONDISCRIMINATION. The ability to carry out genetic testing makes it important to protect people from unfair treatment on the basis of their genetic characteristics (Human Genetics Commisssion 2002). This is a fairly new governing principle in the area of genetics. However, a number of international instruments support this principle, including those developed by the United Nations, the Council of Europe, and a number of European Union political declarations. The 1996 European Convention on Human Rights and Biomedicine prohibits any form of discrimination on the grounds of a person's genetic heritage (Provincial Advisory Committee 2001).

GENETIC SOLIDARITY. Another relatively new principle, intended to recognize the distinctiveness of genetic information, stresses the importance of solidarity in dealing with ethical issues in genetics (Knoppers and Chadwick 2005). The Human Genetics Commission formulates the principle in the following way: "We all share the same basic human genome, although there are individual variations which distinguish us from other people. Most of our genetic characteristics are present in others. This sharing of our genetic constitution not only gives rise to opportunities to help others but it also highlights our common interest in the fruits of medically-based genetic research" (Human Genetics Commission 2002).

3.3 NUTRIGENOMICS TESTING IN THE CLINICAL SETTING

There are a number of principles and obligations covering the obtaining, storage, and disclosure of nutrigenomic information in a clinical setting. It is important to obtain informed consent for all genetic tests and for the production and dissemination of health advice based on a test. Both types of information should be kept confidential, and patients or consumers should know what measure will be taken to keep information secure. Although much of the information about nutrient–gene interactions is gleaned from secondary use of existing data or genetic samples, the secondary use of information generated in a genetic test should be controlled. There are prevailing concerns about the impact that new genetic information can have on other family members, an issue that raises a corollary problem about the appropriateness of testing children and adolescents.

3.3.1 Informed Consent

Although the details of requirements for informed consent vary among jurisdictions, they generally require that the test taker possess adequate information about the nature and purpose of the test, appreciate the foreseeable consequences of taking the test, and consider comparable alternatives to the test in terms of relative benefits and harms. Consent may be expressed or implied, oral or written, depending on the sensitivity of the information in question, the possibility for misunderstanding its nature, and the benefits. The presumption is that the decision to have a nutrigenomic test is an informed decision taken freely. At present, voluntary consent for nutrigenomic testing is not really in question, because the demand for testing is either driven by consumers taking advantage of the few products and services available in the market, or via a physician recommendation. Follow-up on a physician referral implies consent, and if the test is provided privately, the company will generally require clients to sign consent statements.

One traditional criterion for informed consent is that full information relevant to the decision-making process be disclosed and understood. Questions can be raised about the feasibility of fulfilling the information requirement of informed consent (O'Neill 1997). There are several complexities: the quality and complexity of the information generated by nutrigenomic testing, including the probabilistic nature of the results; the uncertain risk–benefit ratio of taking the test; and plausible alternatives to taking the test. Together, the information condition of consent is more difficult to satisfy than in the case of a genetic test for a monogenic disorder. This problem is exacerbated further by the use of tests that simultaneously detect multiple genes associated with different diseases. The challenges that patients or consumers would face in understanding what they are consenting to by taking a test can be mitigated by good pretest counseling, which in the case of private companies, runs the risk of being biased, or in the case of health care practitioners, might be less than ideally informed (Castle 2003).

Informational challenges aside, it is not clear whether consent should be required for genetic tests. Implied consent is normally all that is required for simple, low-risk interventions in which the benefits are believed to greatly outweigh the risks. At present, most nongenetic clinical tests are not subject to consent requirements. This is also the case for some genetic tests, including screening for Down's syndrome (Buchanan et al. 2002). Whether informed consent requirements should apply is complicated further by the fact that according to Henry Greely, "risks to a patient from genetic tests are not the kinds of direct medical risks, such as death or paralysis, that informed consent usually covers" (Greely 1998).

On balance, it may be advisable to require explicit informed consent for nutrigenomic testing. Although the information currently coming from these tests appears to be of a nonsensitive nature and of low informational impact, this could change in the future as scientists uncover more information about gene–health links. Also, a consent requirement, particularly a written consent, would help to ensure that health practitioners provide information about which tests are complicated, with outcomes that are difficult to interpret. There is also a possibility that information from a test may have implications for genetic relatives, an issue we return to later.

3.3.2   Confidentiality

In the medical context, any information that is disclosed to and from the patient, or that emerges during the course of medical care, is protected by strict rules of confidentiality. Genetic information of the sort derived from nutrigenomic tests should receive the same level of protection as that which is given to any other type of health information. Since disclosure of confidential information among medical staff is often necessary for the proper provision of care, consent is not required in standard cases. However, for all disclosures of nutrigenomic information outside the health care context, explicit consent must be sought.

3.3.3   Secondary Information

DNA has been called a "future diary," holding much currently unknown but potentially highly sensitive information about our medical and social prospects (Quaid et al. 2000). This gives rise to a concern about nutrigenomic testing and secondary information. Genes typically have many functions, only some of which are now well understood. Tests done now on a specific gene and its functions may not account for the full range of function the gene possesses—a fact that might be revealed in the future. In this respect, genetic information may, at a later time, reveal knowledge about a person's health risks that were not anticipated at the time of testing. Again, this raises questions about the extent to which full information can be provided or full consent given. People may find that their desire to know certain information about themselves changes as new aspects of their genetic profile come to light.

3.3.4 Families

As one author has put it, "genes are a family affair" (Goldworth 1999). A person who obtains genetic information about his or her risk status by taking a genetic test also gains insights into his or her genetic relatives' risk status (Hayry and Takala 2001). If test results are relevant to family members, should the physician breach patient confidentiality if necessary and inform those members, or should the physician respect patient confidentiality where the patient refuses to share the information and thereby forgo an opportunity to provide potentially useful information to affected members? Or, should the consent requirement be revised to include the consent of all those family members affected by the information?

In the face of these potentially conflicting interests, some affirm the traditional primacy of individual autonomy. One of the main justifications given for this position is the centrality of trust in the health of the patient–physician relationship, a trust that could be undermined by violations of confidentiality (Rhodes 2001). Because genetic information is familial in nature, others have suggested that the traditional focus on individual autonomy needs to be supplemented with a principle of mutuality (Boetzkes 1999; Knoppers 2002). Since the health and welfare of others is at stake, the health practitioner administering the test is thought to have a duty to warn other people under certain conditions. These would include a condition that is serious, where there is a high probability of its occurrence in an identifiable genetic relative, and where there exists the possibility of prevention or treatment (Knoppers 2001).

Whether the health practitioner has a moral obligation to inform other family members of their risks remains uncertain in the case of nutrigenomic testing. According to the criteria above, overriding patient confidentiality would be justifiable only if there is a high probability of serious and irreparable harm done family members if the information is withheld. There must also be some reasonable expectation that disclosure of the information will prevent harm. Nutrigenomic testing does not seem to meet the first condition as long as the predictive value of nutrigenomic tests for particular diseases is low, and our ability to assess the informational value of test results for others is therefore limited. On the other hand, because preventive measures are possible for disease susceptibility far into the future, dietary interventions could be used. Any attempt to reduce known susceptibilities would allow affected family members the opportunity to modify their dietary habits if they had access to this information.

In addition to the ethical obligation of the duty to warn potential at-risk family members, *the right not to know* has been recognized as an equally important right in the realm of genetic testing. This right is based on an awareness that possessing genetic information may be harmful, in that a person should not be coerced to live with the burden of knowledge against his or her wishes, particularly if some information may come with unwelcome obligations (e.g., the obligation to reveal the results to an insurance company). In this respect, the right not to know can be understood as a person's self-determination and a legitimate expression of the basic ethical principle of autonomy. This idea is usually seen in cases of late-onset genetic disorders and where there is no treatment for a genetic condition, such as Alzheimer's disease and breast cancer. The World Health Organization's *Guidelines on Ethical Issues in Medical Genetics and the Provision of Genetic Services* emphasizes the right not to know by saying that "The wish of individuals and families not to know genetic information, including tests results, should be respected, except in testing of newborn babies or children for treatable conditions" (Andorno 2003). Objections to this right refer to the value of solidarity and the responsibility to others. People who choose not to know their genetic status may be putting themselves in the position of being unable to disclose pertinent information to family members.

These concerns suggest that nutrigenomic information may have potential harmful effects. Can this information be a burden to people by significantly compromising their autonomy and psychological integrity? Two ways of assessing the potential harm of nutrigenomic information is by determining the predictability and specificity of the tests in relation to the prospects for treatment and the nonmedical risks and benefits of the tests. In regard to the first criterion, we mentioned earlier that nutrigenomic tests currently focus on determining susceptibilities for low-enetrance genes, which are often implicated in multifactorial pathologies. In the current commercial applications of nutritional genomics, tests sold are often described as susceptibility tests with a wellness focus. Second, in the event that nutrigenomic tests can indicate a person's susceptibility to a disease, a specific and readily available diet would be the known intervention.

3.3.5 Genetic Testing of Children and Adolescents

Genetic tests are performed on children for a number of reasons, including as a response to symptoms that suggest an underlying genetic disorder or for diseases known to run in the family. In nutrigenomic testing, the focus would be on determining a child's susceptibility to future illness and disease, and on the reduction of their risk through early dietary invention, starting as early as infancy (Human Genetics Commission 2005). Some forecast that it will be possible to profile newborns genetically in as little as a decade or two. This raises the question of whether the decision to administer a nutrigenomic test to a child falls within the legitimate range of parental discretion. This important question is raised, we note, in the context of children already being tested for what one might consider the earliest nutrigenomic test: that for PKU.

Parents have wide-ranging discretion when it comes to making medical decisions for their children on the assumption that children are not yet capable of making these decisions themselves and that the parents will act in the best interest of the child. This suggests that genetic testing done to promote the child's welfare should be permitted. On the other hand, some have argued that to preserve the autonomy of the child and the right of the future adult to confidentiality, genetic testing should be prohibited until such time as the child has the ability to make his or her own decision about the testing (Clarke 1998). Several professional societies have published statements advocating a prohibition on the testing of asymptomatic children for late-onset disorders where there is no known intervention or cure (Quaid et al. 2000). The *Inside Information Report*, produced by the UK Human Genetics Commission, recommends exercising great care when considering the testing of children for genetic disorders that cause symptoms only later in life. It concluded that such tests not be performed until the child is old enough to provide his or her own consent to the test (Human Genetics Commission 2002).

Although there is no literature on the topic of nutrigenomic testing and children, there are related discussions. The greatest concern regarding the genetic testing of children is that knowledge of a genetic disposition could lead to emotional damage. Because this information would come during a time of identity development, some fear that children who are found to have increased susceptibility to a disease might suffer from diminished self-esteem. Given their limited understanding of illness, children could view themselves as sick or damaged, and might experience feelings of guilt or self-blame. A second concern is that the results of the test could impair family relationships. Parents might have distorted perceptions about their children's susceptibility. Others contend that the testing of children could lead to their future options being restricted, due to a sense of fatalism about what may lie ahead. A fourth set of concerns has to do with the risk of future uses of insurance to risk-rate people, and potential employment discrimination against children found to be more susceptible to certain conditions. Finally, critics of testing point out that parents are effectively usurping the right of the future adult to make the choice about testing thereby violating his or her autonomy.

Most of these harms have been discussed in the context of testing asymptomatic children for adult-onset disorders. In the context of serious genetic diseases, where there is no predictable time of appearance or of severity, and no effective therapy, it is important to take these potential harms seriously. However, many of these concerns would not appear to apply in the nutrigenomics context, due to differences in the nature of the tests in question. Nutrigenomic testing is discretionary testing, and is of low predictive value for future illness. The potential for immediate harm is therefore lower in these regards. Where there are known interventions, the long-term health benefits could offset any negative harm associated with being tested. In addition to the long-term benefits to the health of the child derived from the test result information, there could also be immediate benefits. Parents have the right and perhaps an obligation to make the best possible health care decisions for their children. On these grounds, administering nutrigenomic tests to children may be morally permissible, but it is a delicate balance that is being sought.

On the other hand, the scientific validity of nutrigenomic tests, as well as the efficacy of dietary advice recommended for improving health outcomes, has yet to be broadly confirmed. There is also a concern, based on documented cases, that other types of susceptibility screening have harmed the health of children. For example, there is some evidence that children have been harmed by dietary restrictions after cholesterol screening, leading some to suggest that similar problems could arise after genetic testing for susceptibility to a broader range of common diseases (Clarke 1998). Most nutrigenomic dietary advice is developed for adults rather than for children. The U.S. Preventive Services Task Force also points out that little is known about effective dietary counseling for children in a clinical setting. Most studies on nutritional interventions for children have taken place in nonclinical settings. Finally, we have to weigh the quality of the advice gained through nutrigenomic testing against the advice already available to the public about how best to meet children's nutritional needs. It may be that these tests add little to the information already available to parents, a situation which, if it were to change, would speak in favor of limited use of pediatric nutrigenomic tests.

A number of bodies have laid out guidelines to assist in making determinations about when it is appropriate to test children genetically, and this may be of help in the nutrigenomic context. For example, the U.S. Genetic Privacy Act has set out criteria for determining when it is suitable to perform asymptomatic genetic tests on minors. The act suggests a ban on testing unless (1) there is an effective intervention that will prevent or delay the onset or ameliorate the severity of the disease, and (2) the intervention must be initiated before the age of 16 to be effective (Laurie 2001). The rationale for this advice is twofold. First, there is a concern that the information could be used to stigmatize or discriminate against the child now and in the future. There is also a presumption that "a child's genetic status is the child's private genetic information and should not be determined or disclosed without some compelling reason" (Laurie 2001). To meet these criteria, we would have to know that there was a significantly improved prognosis for the child as a result of early nutrigenomic screening, and that effective dietary interventions were available. Although some progress has been made on setting international standards for when it is appropriate for children to receive genetic tests, standards are expected to remain controversial and to evolve (Dalby 1995; Freyer 1995, 2006).

3.4 USE OF NUTRIGENOMICS INFORMATION FOR RESEARCH

With an increase in knowledge about the genetic basis of common diseases and individual responses to drugs and nutrients, many more persons will undergo DNA testing as part of ordinary medical treatment and prevention. Repositories of both DNA samples and DNA-based information will become part of public and private health care systems. For patients and consumers to accept nutrigenomics testing and nutrigenomics research, a clearly defined set of rights and protections for genetic material and information needs to be established. Those rights and protections should specify when and how DNA samples are obtained and used for either clinical or research purposes and how they are stored. Whether public or private, nonprofit or commercial, most ethical advisory commissions agree that the patient's autonomy, privacy, and confidentiality should be protected when DNA material has been collected or will be collected in the future. Differences arise in how best to implement these principles in practice. Other major concerns include the withdrawal of material, sharing of benefits, researcher access, and secondary use of information by insurers and employers.

National biobanks and databases have been established in Iceland, Sweden, Estonia, Tobago, and the United Kingdom. DNA samples and test results derived specifically from nutrigenomic testing are not currently gathered and stored in these government-sponsored banks, but this could change. In the interim, some private databanks do appear to have an interest in the nutrigenomic information being gathered. For example, the commercial owners of the Icelandic Healthcare Database had originally promised Icelanders benefits such as jobs and free drugs for their health system in exchange for exclusive rights to the information (Robertson 2003). With consumer consent, the company proposed to sell the information to pharmaceutical companies that would find it beneficial in their drug development processes.

Storage and use of samples and information for future research raises important challenges for meeting consent, privacy, and confidentiality requirements. The fast pace of scientific research makes it difficult, if not impossible, to predict the many uses to which genetic information might eventually be put. This raises questions about how practical it is for people to give meaningful informed consent as it is traditionally understood. Privacy and confidentiality present special challenges in the research setting as well. On the one hand, anonymizing research samples and data lessens the potential for violations of privacy and confidentiality, thereby reducing concerns about consent. On the other hand, sample and data identification is desirable for both individual and social reasons. People may want to be notified of new research results that have medical implications for themselves and their families (Caulfield et al. 2003).

Researchers may need to link data from various sources with identifying markers. It is desirable that "genetic epidemiologists [are] able to correlate clinical, outcome and utilization data and, in longitudinal studies, to update data over time" (Anderlik and Rothstein 2001). Anonymizing samples and data could have the effect of impeding socially beneficial scientific research and development. Depending on the research in question, this could have adverse consequences for the health of appreciable numbers of people. Additional areas in need of exploration include questions about ownership of samples, data and resulting research, storage security measures, time limits for storage, and who should have access to samples and data and for what purposes.

It should also be pointed out that genetic testing can often confirm, or provide finer-grain understanding of, disease susceptibility learned through pedigree analysis. Pedigrees have been described as the "lifeblood of research in human genetics [because] they provide a shorthand description of the genetic and social relationships, and they allow modes of transmission of heritable traits or patterns of inheritance of haplotype to be assessed at a glance" (Byers and Ashkenas 1998). Pedigrees might be particularly useful in nutrigenomics because they capture patterns of gene–gene as well as gene–environment interactions that may be relevant to nutrigenomics (Hunt et al. 2003). Like genetic information and biological samples, there is disagreement about the extent to which pedigree information should be protected based on the potential for harm should the information be made public. The prospect, for example, of pedigree information about a family history of Alzheimer's disease being made public is a concern, partly because academic journals have inconsistent standards that are administered inconsistently (Botkin et al. 1998; Nisker and Daar 2006).

Robert Cook-Deegan has argued that although a person's pedigree information is relevant to protecting that person from harm, it is an open question about the status of the other people named in the pedigree: "The real question is when informed consent is necessary, as well as when privacy protections can be relied on even for those from whom informed consent is not explicit. This line of argument creates a new class of people to whom duties of confidentiality are owed but who could be handled under the regulations without making them 'human subjects.' But if they are not subject, then what are they? This calls for serious public debate and formal policy making" (Cook-Deegan 2001). As Cook-Deegan points out, what counts as private information is not always determinate and it raises questions about whether pedigrees inappropriately disclose information about, for example, identifiable populations (Foster and Freeman 1998).

Fortunately, these issues have not escaped the attention of researchers in the field. John Mathers, who leads one of the work packages for the European Nutrigenomics Organisation (NuGO), has remarked that "there is some wariness among both the public and professionals about the acquisition, storage and sharing of any genetic information about identifiable individuals that must be addressed if this research approach is to fulfill its promise" (Mathers 2004). In 2005, NuGO convened a workshop of advisors drawn internationally to provide advice on a draft set of research ethics guidelines that are expected in 2006 (Bergmann et al. 2006).

3.5 USE OF NUTRIGENOMICS INFORMATION BY PRIVATE THIRD PARTIES

As more is learned about individual genetic susceptibility to disease, the information garnered from genetic tests will become increasingly attractive to outside parties who stand to gain from it. We look at disclosure of genetic information to third parties and, in particular, at concerns about the potential for private third-party discrimination based on nutrigenomic test information. We focus on the use of this information primarily in the contexts of employment and health insurance, but also identify issues for further consideration in the areas of life, disability, and long-term care insurance. There are concerns that private parties may seek to use genetic information about a person to make unfair decisions that affect that person's welfare.

The extent to which this is valid concern depends in part on the jurisdiction and its health care system as well as the country's legal and political structures. In Canada and the United Kingdom, universal health care coverage makes concerns about standard health insurance uses of genetic information to risk-rate clients less urgent. In the United States, where private insurance plays a greater role, concern is greater. However, in the United Kingdom, where one's ability to secure life insurance has implications for securing home and automobile loans, concerns about unfair risk rating in this market are greater than in other jurisdictions.

Concern about genetic discrimination by third parties runs high among the public in most countries where genetic tests are in use. A 1995 survey found that over 85 percent of respondents described themselves as "somewhat to very concerned" about access to and use of genetic information by insurers and employers. In the United States, public fears of discrimination in health insurance and employment are often reported as major deterrents to genetic testing (Greely 1998). In a 1997 survey, two-thirds of the respondents reported that they would not undergo genetic testing if they thought that health insurers and employers would have access to the results (Hall and Rich 2000a). Insurers and employers are currently allowed to differentiate among people on a range of grounds in the process of risk-rating clients. The question is whether it is inherently inappropriate to risk-rate people on genetic grounds. The overriding fear is that using genetic information could lead to the loss of access to affordable health care, to employment, and to the ability to provide for dependants in case of accident or death. Important ethical principles are at stake here, including consent, privacy, confidentiality, and genetic nondiscrimination (Human Genetics Commission 2002).

3.5.1 Insurance

Health, life, disability, and long-term care insurers already engage in risk classification on the basis of such factors as age, individual and family medical history, current and past health status, occupation, serum cholesterol, and alcohol and tobacco use. Technology is allowing insurance companies to categorize people and their risks ever more precisely, and genetic testing adds one more level of detail. Genetic testing may provide a way of obtaining more accurate assessments of a person's risk of morbidity and mortality. This information could be used to discriminate more finely between the risk levels of different individuals, and this could alter the availability and cost of health, life, and other types of insurance. In the private health insurance context, the issue revolves around medical underwriting and preexisting condition limitations. As Greely points out, "insurers and health maintenance organizations medically underwrite consumers when they decide whether to issue coverage and at what cost, based in part on information about the consumers' predicted future health" (Greely 1998). This may be reason enough to consider ethical and legal approaches to the problem despite the fact that actual genetic is not currently a problem (Greely 2005).

The possibility that people might be denied insurance because of their genetic background may add more fuel to the debate about how personalized the risk analysis and premium setting should be. According to the British Data Protection Act, a duty of fair processing requires that the personal data used must not be excessive and that it be relevant to the actuarial process. This could open the door to the use of genetic information if it meets these conditions (O'Neill 1997). It is conceivable that one day genetic tests may be required as a condition for insurance eligibility. This potentially coercive collection of information could be construed a violation of consent requirements and of the value of autonomy and respect for the person. Mandatory testing would also amount to an invasion of privacy by allowing the insurance industry forced access to applicants' bodies. The result could be the loss of essential goods, in this case those related to health care (Anderlik and Rothstein 2001).

Would differentiating between people on the basis of a person's genetic makeup be an ethically permissible practice? Some have taken the position that genetic characteristics of any type should be treated like other characteristics over which a person has little to no control, such as race, gender, and age (Cook 1999). According to the Human Genetics Commission, "An individual's genetic information is arguably different from other 'relevant information' that is a matter of lifestyle choice, such as dangerous sports or driving a sports car" (Human Genetics Commission 2002). Furthermore, to the extent that certain genetic mutations have a higher rate of occurrence in some groups than others, there is a risk of group-based stigmatization and unfair use of genetic information, practices that are globally condemned (O'Neill 1997). A middle position might hold that the permissibility of using genetic information could be determined, in part, by the importance of the issues at stake. Based on this approach, genetic information use could be banned for health insurance, but perhaps allowed in life insurance.

In many countries, life insurance is thought to be less essential than health insurance. In contrast to health insurers, life and disability insurers have a greater stake in making use of predictive genetic information that could alert them to a person's probable health status many years into the future (Hall and Rich 2000b). As well, "these policies tend to be sold in the individual market which heightens an insurers' interest in individual health risks" (Anderlik and Rothstein 2001). From the consumer's viewpoint, life insurance is often an important means of financial planning. In some contexts, the acquisition of life insurance is tied to the achievement of other important goods. Medically underwritten life insurance can be important for acquiring home or automobile loans.

As the median age of the population increases, the demand for long-term care will increase. With medical costs rising, it is expected that at least some countries that publicly fund these programs will try to shift long-term care costs to the private sector. Like life insurance, private long-term care insurance is sold in the individual market, so the same risks apply. Long-term care insurers are most concerned about a person's prospects for morbidity without mortality, since this is where their payouts will be greatest. They have a strong economic interest in any information that can be used to predict risk of future illness, including that derived from nutrigenomic tests where diet–gene–health outcomes have been well established scientifically.

At present, there are few limits on the use of genetic information in the area of life, disability, and long-term care insurance. No state law prohibits a life insurance company from asking an applicant to take a genetic test (but when a test has been done, isn't there a requirement in those positioned to divulge this information and the results of the test to stop people with poor prognoses from overinsuring themselves?). While a number of states in the United States have enacted legislation aimed at life insurance, the protections afforded applicants are fairly minimal. "Typically these laws require only that informed consent be obtained prior to the performance of genetic testing and/or that any use of genetic information in medical underwriting meet standards of actuarial fairness" (Anderlik and Rothstein 2001). Other approaches, favored by countries such as Canada, the United Kingdom, and the Netherlands, prohibit the use of genetic information for life insurance policies below a certain dollar amount. Legislators have been even slower to act in the area of long-term care insurance. Only two states in the United States regulate the use of genetic information in the issuance of long-term care insurance; and the United Kingdom has a moratorium on the use of genetic information for insurance (except for a few, primarily neurodegenerative conditions). We identify this as an area of future research that will become more important as our ability to make predictions about susceptibility to future illness improves.

The Canadian Genetics and Life Insurance Task Force has agreed to examine two options for dealing with genetic information and insurance. One idea is a five-year moratorium on the use of genetic tests results (excluding family history) for insurance coverage. The other is to create an independent standing body (as in the United Kingdom), which includes consumers, government, clinicians, industry, and researchers, for ongoing review of criteria on the reliability of genetic information for underwriting purposes (Knoppers et al. 2004).

A number of harms could result from the misuse of nutrigenomic information in the context of insurance. The first is a perceived danger that discrimination might result from decision-making made on the basis of faulty or incomplete data, misunderstandings about the genetic science involved, or misinterpretations of the predictive value of genetic test results for morbidity and mortality. Public fear of misuse of information could discourage people from undergoing beneficial genetic testing, where health-promoting interventions are relatively easy. In the case of nutrigenomic testing, a positive test result may make it possible for a person to take specific steps, in the form of simple dietary interventions, to alter their level of risk for certain conditions. To the extent that people fear abuse of test results and avoid this technology, they may lose out on opportunities for preventive measures that could improve their health. Fears generated by concerns about genetic discrimination could also contribute to public reluctance to participate in national research programs that may, in turn, impede developments in the prevention and treatment of disease. Finally, fears about possible discrimination could accelerate the move away from test-taking provided in the health care setting and toward private-based testing, possibly leaving test-takers with inadequate information and professional support.

Despite these potential harms, public fears about genetic discrimination appear to be unfounded. There is a growing consensus that much of the evidence regarding genetic discrimination widely reported in the past has been distorted or overblown because it was based on anecdotal reporting or derived from studies with serious methodological flaws (Anderlik and Rothstein 2001). A study reported in 2000 from the United States found almost no well-documented cases of health insurers either asking for or using genetic tests in their underwriting decisions where the person had no symptoms (Hall and Rich 2000b). Other American studies have found that insurers already cover people who have serious genetic disorders or dispositions to disease (Anderlik and Rothstein 2001). Insurance companies in the United Kingdom do not currently request that people take genetic tests, and the industry claims that this will remain its policy for the foreseeable future (Cook 1999). Industry insiders cite a number of reasons for this exclusion, including a belief that it would be unethical to penalize people for genetic defects that have not yet manifested as well as concerns about political and public relations backlashes that would probably result from using this type of information (Hall and Rich 2000b).

There are also a number of reasons why genetics is still of little use to insurers. Genetic information, including that from nutrigenomics testing, is not sufficiently predictive of future ill health and is still considered irrelevant in insurance decision-making. Like other forms of genetic information, nutrigenomic information is unreliable on a number of grounds, including a lack of knowledge about the degree of association between the possession of certain genes and the likelihood of disease; the time of onset and the severity of disease associated with the genes; the effectiveness of dietary interventions which may modify the effects of these genes; and the extent to which other genetic, behavioral, and environmental factors play a role in the effects of these genes and their impact on health (O'Neill 1997).

The fact that there is currently little evidence of discrimination in health insurance practices does not mean that this will continue to be the case in the future. These tests have come into clinical use very recently, and the rate of uptake is expected to grow rapidly in the future. Our scientific understanding may develop to the point that sound actuarial data based on genetic test results can be generated. It may become possible for insurers to claim rational, scientifically sound, and empirically supported grounds for making risk ratings on the basis of individual (and perhaps group) genetic profiles. In one set of interviews, four of six industry members said that if more genetic information becomes available and if the predictive nature of these data becomes sufficiently precise, it probably would become much more relevant to medical underwriting (Hall and Rich 2000b).

Governments, policymakers, the public, and industry need to strike an appropriate balance between actuarial fairness and social fairness in insurance practices. There are two often-competing values underlying insurance practices: solidarity and equity. Solidarity requires social risk-spreading as a way of ensuring that the benefits and costs of the insurance are shared by the population as a whole, including its subgroups. This ensures coverage of essential health care needs for all citizens. Equity involves individual risk diversification and requires that the contributions of individuals be in line with their known levels of risk. If people bring extra risk into the insured pool, they ought to bear proportionate additional costs rather than have those costs passed on to other policyholders (Cook 1999). According to proponents of actuarial fairness who argue on the basis on equity, failure to make use of this information would mean that insurers would be forced to charge the same rate to insured persons who have different expected costs. This type of insurance pricing, some claim, would be unfairly discriminatory to the group of persons who have lower expected costs. Others claim that we need to look at the larger moral and social consequences of permitting the use of genetic information for insurance purposes (O'Neill 1997).

3.5.2 Employment

Employers can use genetic testing for several purposes. They can be used as screens to try to predict whether prospective or current employees have genetically determined traits likely to make them more susceptible to certain workplace hazards, such as chemicals. There is a notorious case in which Burlington Northern Sante Fe Railway wanted to test employees filing for carpal tunnel syndrome–related disability claims. The test was for hereditary neuropathy to recurrent pressure-sensitive palsies, which is an inherited disorder. The railway stopped testing employees when the U.S. Equal Employment Opportunity Commission filed a suit against them (Friedrich 2002).

Another type of testing is for genetic susceptibility to diseases that might make prospective and current employees less productive in the workplace, more likely to take sick leave, more costly in terms of health insurance, and more likely to retire early due to illness. Nutrigenomic testing would fall under this category. There is a close connection between health insurance and employment uses of medical information in some countries, where health care costs are cited as a major financial concerns for employers (Cooper 2002). For example, in the United States, where many employers fund their employees' health care, there is an economic incentive to seek genetic information about job applicants and employees as a way of controlling expenditures. Genetic testing can also be relevant to employment-related benefits such as pensions and annuities. From the point of view of the employee, the concern is that genetic test information could be used unfairly in hiring, promotion, termination, and other employment decisions, as well as in decisions about health and other benefit packages.

As with insurance discrimination, employment discrimination on the basis of genetic testing is more of theoretical than practical interest. At present, there is little risk that employers will conduct genetic testing themselves, because it is not cost-effective, and the resulting information is still not very precise (Cooper 2002). According to the *Inside Information Report*, there is no evidence that employers in the United Kingdom are systematically using

A widely publicized case in 2001 involved a suit against the Burlington Northern Santa Fe Railway, which tested workers who claimed compensation for carpal tunnel syndrome for a rare genetic condition that may have predisposed them to the syndrome.

Professional athletes, given the direct connection between their physical well-being and on-the-job performance, are prime candidates for work-related genetic discrimination. Preexisting health conditions have already been used to ban athletes from competing. Athletes who are diagnosed with hypertrophic cardiomyopathy (HCM), for example, have been prevented from competing in international competitions. Some of the genes associated with HCM are known, but a wholly reliable test is not yet available. Still, the hope is that a test might be made available to diagnose the condition, and potentially, to rate the risk of those likely to suffer sudden death from those who are at low risk (Spinney 2004). Nevertheless, the Chicago Bulls player Eddy Curry was asked by team management to have a genetic test for HCM, on the basis of his having an enlarged heart and irregular heartbeat. For a short time, the team would not allow Curry to play without the test, which Curry refused to take, threatening legal action. Curry was eventually allowed to return to regular play without taking the genetic test.

genetic test results to recruit or deny people employment, nor are they using such tests as part of workplace health programs (Human Genetics Commission 2002). In the United States, a study by the American Management Association showed that only 0.5 percent of employers were requiring genetic tests, some of which were performed in compliance with directions from the U.S. Occupational Safety and Health Administration. According to a second study done in 2000, only 0.004 percent of employers were engaged in genetic testing, where "tests" were defined as those that analyze a person's genes (Cooper 2002).

We need to ask whether it would be ethically permissible for employers to take an interest in accessing genetic information. The issue is usually cast as a question about how to balance the employer's right to know information that is relevant to the workplace with the employee's right to privacy (Pagnatarro 2001). In the case of nutrigenomic testing, it is not clear that employers should be able to request these tests or gain access to test result information, since the information is not employment related. On the other hand, employers do have financial interests at stake in employee health. They could argue that there is a legitimate business reason for access to this type of genetic information. On this question, the UK Human Genetics Commission (HGC) concluded that is inappropriate to use genetic testing unless there are clear occupational health benefits, a position that would rule out the use of nutrigenomic tests (Human Genetics Commission 2002). In contrast to this, survey results gathered by the HGC from the broader community indicated somewhat different results. Whereas some felt that it was inappropriate to use genetic test results to try and predict who might be absent from work more frequently, there was some support on the part of others for using genetic testing to help predict sickness absenteeism or early retirement.

3.5.3   Legal and Social Responses to Fears of Discrimination

Despite the lack of evidence of actual discriminatory practices in insurance and employment, the introduction of antidiscrimination legislation has become a popular method of regulation in some countries, probably in response to the heightened levels of public concern.

In 2005, the U.S. Senate passed the Genetic Information Nondiscrimination Act, which sought to bar companies from using genetic information to deny health coverage or employment, and a Presidential Order banned genetic discrimination in federal employment. More than 40 states have enacted genetic nondiscrimination laws, which typically prohibit insurers from requiring genetic tests or test information as a condition of insurance, from charging nonstandard premiums or from restricting the scope of conditions covered on the basis of known susceptibility tests results (*NIH News* 2004). However, some states permit the collection of employee genetic information where that information is jobrelated or is in connection with "a bona fide employee welfare or benefit program" (Anderlik and Rothstein 2001). A range of additional existing legal instruments has been identified as offering possible protections against workplace discrimination, depending on how they are interpreted. These include the Americans with Disabilities Act, Title VII of the Civil Rights Act of 1964, the Occupational Safety and Health Act, the U.S. Constitution, and a variety of state laws (Pagnatarro 2001; Cooper 2002). The U.S. Health and Insurance Portability and Accountability Act precludes the use of genetic information to establish a preexisting condition in the absence of a diagnosis.

According to Anderlik and Rothstein, "Internationally, most countries regulate the collection, storage and use of genetic information, including use by third parties, through omnibus data protection legislation" (Anderlik and Rothstein 2001). At the international level, the UNESCO *Universal Declaration on the Human Genome and Human Rights* states that "no one shall be subjected to discrimination based on genetic characteristics that is intended to infringe or has the effect of infringing human rights, fundamental freedoms and human dignity" (Godard 2003). In Europe, the UK Human Genetics Commission has recommended that employers not be able to request that employees take a genetic test as a condition of employment (Human Genetics Commission 2002). The British government, in cooperation with the Association of British Insurers, agreed to a five-year moratorium, ending in 2006, on the use of genetic information for insurance decision purposes. The government asked the Human Genetics Commission to undertake a review of genetic discrimination in employment in 2005. In 2006, the HGC concluded that discrimination was still a potential threat and should be closely monitored.

The Netherlands, France, Denmark, Luxemburg, Austria, Norway, and Belgium have introduced measures that either prohibit or restrict the use of personal genetic information in insurance and employment (Hall and Rich 2000b). A number of governments want to allow further scientific research into the nature of genetic predispositions to ill health, and to provide time for wide consultation and careful long-term policymaking. In 2003, a major Australian report proposed creating a national commission on human genetics and recommended that federal antidiscrimination legislation be amended to "prohibit unlawful discrimination based on a person's real or perceived genetic status."

A number of genetic counselors and other health care providers have adopted some informal practices aimed at reducing the potential for genetic discrimination. These include medical descriptions of test-taking that mask their genetic nature, placing genetic test results in separate "shadow" files rather than in medical records and encouraging clients to pay for tests out of pocket rather than seeking reimbursement from insurance companies (Hall and Rich 2001a).

3.6 CONCLUSION

Nutrigenomics requires the collection of biological samples, from which genetic information is derived. The potential to link genetic information and samples to personal information, including other medical information, persists in research and clinical contexts, a fact that raises the ethical challenges associated with most genetic tests. Since nutrigenomic tests are currently seen as a form of discretionary testing of a relatively nonsensitive nature, and having mild to moderate informational impact, the information generated by these tests is not currently considered to be exceptionally harmful in the sense that the threat of immediate and significant harm is not imminent. This does not mean, however, that people undergoing tests, or conducting testing, should be any less vigilant about controlling nutrigenomic information. Given that nutrigenomics is a growing area of scientific research, the future potential impact of information mismanagement might be greater. Therefore, we recommend that the highest standards for information management be adopted in nutrigenomics to set the bar at a high level from the outset, rather than having to recover lost ground in the future.

Those who offer nutrigenomics tests should follow applicable regulatory procedures and professional standards for obtaining informed consent, providing access to appropriate counseling, and protecting confidentiality and privacy. Information from nutrigenomics tests should not be used in any way that causes discrimination against individuals, families, children, or vulnerable groups in either clinical or nonclinical contexts. Finally, nutrigenomics information should not be used by third parties, such as employers and insurers, unfairly to discriminate against people.

REFERENCES

Anderlik, M. R., and M. A. Rothstein. 2001. Privacy and confidentiality of genetic information: What rules for the new science? *Annual Review of Genomics and Human Genetics* 2:401–433.

Andorno, A. 2003. The right not to know: an autonomy based approach. *Journal of Medical Ethics* 30:435–440.

Bergmann, M. M., M. Bodzioch, M. L. Bonet, C. Defoort, G. Lietz, and J. M. Mathers. 2006. Bioethics in human nutrigenomics research: European Nutrigenomics Organisation (NuGO) workshop report. *British Journal of Nutrition* 95:1024–1027.

Boetzkes, E. 1999. Genetic knowledge and third-party interests. *Cambridge Quarterly: Healthcare Ethics* 8:386–392.

Botkin, J. R., W. M. McMahon, K. R. Smith, and J. E. Nash. 1998. Privacy and confidentiality in the publication of pedigrees: a survey of investigators and biomedical journals. *Journal of the American Medical Association* 279:1808–1812.

Buchanan, A., A. Califano, J. Kahn, E. McPherson, J. Robertson, and B. Brody. 2002. Pharmacogenetics: ethical issues and policy options. *Kennedy Institute Ethics Journal* 12:1–15.

Byers, P. H., and J. Ashkenas. 1998. Pedigrees: publish, or perish the thought? *American Journal of Human Genetics* 63:678–681.

Castle, D. 2003. Clinical challenges posed by new biotechnology: the case of nutrigenomics. *Postgraduate Medical Journal* 79:65–66.

Caulfield, T., R. E. G. Upshur, and A. S. Daar. 2003. DNA databanks and consent: a suggested policy option involving an authorization model. *BioMed Central Medical Ethics* 4:1.

Clarke, A. 1998. *The Genetic Testing of Children.* Oxford: BIOS Scientific Publishers.

Cook, E. D. 1999. Genetics and the British insurance industry. *Journal of Medical Ethics* 25:157–162.

Cook-Deegan, R. 2001. Privacy, families and human subject protections: some lessons from pedigree research. *Journal of Continuing Education in the Health Profession* 21:224–237.

Cooper, Christine Godsil. 2002. Your genes or your job: genetic testing in the workplace. *Employee Rights Quarterly* 3:1–12.

Council of Europe. 1997. *Convention for the Protection of Human Rights and Dignity of the Human Being with Regard to the Application of Biology and Medicine: Convention on Human Rights and Biomedicine.*

Dalby, S. 1995. Genetic Interest Group (GIG) response to the UK Clinical Genetics Society report "The genetic testing of children." *Journal of Medical Genetics* 32:490–494.

Foster, M. W., and W. L. Freeman. 1998. Naming names in human genetic variation research. *Genome Research* 8:755–757.

Freyer, A. 1995. Genetic testing of children. *Archives of Disease in Childhood* 73:97–99.

———. 2006. Inappropriate genetic testing of children. *Archives of Disease in Childhood* 83:283–285.

Friedrich, M. J. 2002. Preserving privacy, preventing discrimination becomes the province of genetics experts. *Journal of the American Medical Association* 288:815–819.

Godard, B., S. Raeburn, M. Pembrey, M. Bobrow, P. Farndon, and S. Aymé. 2003. Genetic information and testing in insurance and employment: technical, social and ethical issues. *European Journal of Human Genetics* 11:S123–S142.

Goldworth, A. 1999. Informed consent in the genetic age. *Cambridge Quarterly: Healthcare Ethics* 8:393–400.

Greely, H. T. 1998. Legal, ethical and social issues in human genome research. *Annual Review of Anthropology* 27:473–502.

———. 2005. Banning genetic discrimination. *New England Journal of Medicine* 353:865–867.

Hall, M. A., and S. S. Rich. 2000a. Patients' fear of genetic discrimination by health insurers: the impact of legal protections. *Genetics in Medicine* 2:214–221.

———. 2000b. Laws restricting health insurers' use of genetic information: impact on genetic discrimination. *American Journal of Human Genetics* 66:293–307.

Hayry, M., and T. Takala. 2001. Genetic information, rights, and autonomy. *Theoretical Medical Bioethics* 22:403–414.

Human Genetics Commision. 2002. *Inside Information: Balancing Interests in the Use of Personal Genetic Data*. London: HGC.

———. 2003. *Genes Direct: Ensuring the Effective Oversight of Genetic Tests Supplied Directly to the Public*. London: HGC.

———. 2005. *Profiling the Newborn: A Prospective Gene Technology?* Report from a Joint Working Group of the Human Genetics Commission and the UK National Screening Committee. London: HGC.

Hunt, S. C., M. Gwinn, and T. D. Adams. 2003. Family history assessment: strategies for prevention of cardiovascular disease. *American Journal of Preventive Medicine* 24:136–142.

Knoppers, B. M. 2002. Genetic information and the family: Are we our brother's keeper? *Trends in Biotechnology* 20:85–86.

Knoppers, B. M., and R. Chadwick. 2005. Human genetic research: emerging trends in ethics. *Nature Reviews: Genetics* 6:75–79.

Knoppers, B. M., T. Lemmens, B. Godard, and Y. Joly. 2004. Genetics and life insurance in Canada. *Canadian Medical Association Journal* 170:1–3.

Laurie, G. T. 2001. Challenging medical–legal norms: the role of autonomy, confidentiality, and privacy in protecting individual and familial group rights in genetic information. *Journal of Legal Medicine* 22:1–54.

Mathers, J. 2004. Symposium on "how and why measure individual variability," chairman's introduction: What can we expect to learn from genomics? *Proceedings of the Nutrition Society* 63:1–4.

NIH News. 2004. International consortium completes human genome project. U.S. Department of Health and Human Services. Retrieved February 2, 2006, from `www.nih.gov/news/pr/apr2003/nhgri-14.htm`.

Nisker, J., and A. S. Daar. 2006. Moral presentation of genetics-based narratives for public understanding of genetic science and its implications. *Public Understanding of Science* 15:113–123.

O'Neill, O. 1997. Genetic information and insurance: some ethical issues. *Philosophical Transactions of the Royal Society of London, B: Biological Science* 352:1087–1093.

Pagnatarro, M. A. 2001. Genetic discrimination and the workplace: employee's rght to privacy v. employer's need to know. *American Business Law Journal* 39:139–185.

Provincial Advisory Committee on New Predictive Genetic Technologies. 2001. *Genetic Services in Ontario: Mapping the Future.*

Quaid, K. A., M. Powers, M. A. Bobinski, M. Anderlik, and R. D. Pentz. 2000. Genetic information, ethics, privacy, and confidentiality: overview. In Vol. 1 of *The Encyclopedia of Ethical, Legal, and Policy Issues in Biotechnology*, edited by T. E. Murray, and M. J. Mehlman. New York: Wiley.

Rachels, J. 1975. Why privacy is important. *Philosophy and Public Affairs* 4:323–333.

Rhodes, R. 2001. Confidentiality, genetic information, and the physician–patient relationship. *American Journal of Bioethics* 1:26–28.

Robertson, J. A. 2003. Ethical and legal issues in genetic biobanking. In *Population and Genetics: Legal and Socio-ethical Perspectives*, edited by B. M. Knoppers. Leiden, The Netherlands: Martinus Nijhoff.

Spinney, L. 2004. Heart-stopping action. *Nature.* Retrieved June 20, 2006, from `www.nature.com/news/2004/040802/pf/430606a_pf.html`.

UNESCO. 1997. Universal declaration on the human genome and human rights. Retrieved November 11, 1997, from `http://portal.unesco.org/en/ev.php-URL_ID =13177&URL_DO=DO_TOPIC&URL_SECTION=201.html`.

4

ALTERNATIVES FOR NUTRIGENOMIC SERVICE DELIVERY

4.1 INTRODUCTION

With the recent completion of the Human Genome Project, there is reason to believe that genetic tests and services will increasingly become available for a greater number of medical conditions. This is a safe prediction to make, despite the hype about the impact of genetics (Bubela and Caulfield 2004) and doubts about the uptake of genomics and genetics into clinical practice and health care (Cooper and Psaty 2003). As the number of genetic tests continues to rise and as genetic information use in health care grows more prevalent, ensuring that genetic services meet the health needs of patients and consumers becomes more complex. Genetic services, which we define here as genetic testing and genetic counseling, will not only broaden in scope and number of services, but the associated infrastructure for service delivery will become more varied and complex over time.

Nutrigenomic service delivery is currently based on three main sources of information which can be used in different ways and in different combinations: a lifestyle questionnaire, which can be separate from a dietary intake assessment, and a genetic test. Each of these elements can be provided by the same or a different company or health care practitioner, and pre- and

Science, Society, and the Supermarket: The Opportunities and Challenges of Nutrigenomics,
By David Castle, Cheryl Cline, Abdallah S. Daar, Charoula Tsamis, and Peter A. Singer
Copyright © 2007 John Wiley & Sons, Inc.

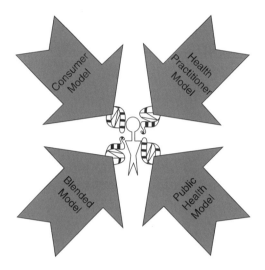

Figure 4.1. Nutrigenetic services are offered in a variety of ways.

post-testing counseling can be provided separate from the administration of questionnaires and tests. Even a cursory glance at service delivery reveals that the delivery of nutritional genomics has the potential for great complexity. Service delivery becomes all the more complex when considerations of quality control; authority for certain types of activities, such as administering tests; or dietary counseling and the types of health claims that can be made, are taken into consideration.

Perhaps the key element to be considered in any discussion of service delivery alternatives is who the nutrigenomics client is. Commercial activity and consumer demand are likely to grow in the wake of continuing scientific advances, but it is unclear whether the dominant trend will be personalized nutrition sought in the marketplace, services offered in a medical context in parallel with conventional medical advice, or some combination of the two. Nutrigenomics research is expected to produce new information about the genetic links to common medical conditions which when coupled with new findings about effective lifestyle interventions, including dietary choices, for those at increased risk should lead to a greater range of nutrigenomic services. There may not be clients for every nutrigenomics service that may be offered. The market for nutrigenomics may become structured idiosyncratically by demographics, succession of nutrigenomic service types, and other factors.

Although the nutrigenomic service market is currently small, a number of drivers are expected to propel its expansion in the near future. Consumer research conducted by groups such as the Institute for the Future strongly suggests that over the next decade, educated, proactive consumers will seek ever more kinds of nutritional and health information sources (Institute for the Future 2003). A wave of early adopters are predicted to increase demand for personalized health and dietary information in contexts where access to direct-to-consumer tests are available. In other contexts, early adopters might find access to nutrigenomic services in a medical context, be it private or public. The fact that access to nutrigenomics will depend on context raises the issue of regulation, which is the subject of Chapter 5. Suffice it to say for now that the size of the nutrigenomic market will always be partially determined by the regulatory environment in which the services are offered. Regulations specific to nutrigenomics have yet, at the time of writing, to be developed in any jurisdiction. The development of regulations could provide the regulatory certainty needed to stabilize the business environment, and a rapid growth in the numbers of companies offering these services could follow.

No matter who the client is, it remains the case that the introduction and expansion of nutrigenomic services raises important questions about how these tests should be administered, by whom, and with what accompanying supports. Genetic services offered to the public ought to be offered in conjunction with genetic counseling because counseling may be, in many cases, pivotal in ensuring the effective use of genetic tests and the success of the interpretation of nutrigenomic information. In a recent literature review, Wang et al. categorize three overarching areas that reflect the main goals of genetic services: (1) to educate and inform patients and consumers of the

Institute for the Future

The California-based Institute for the Future (www.iftf.org) is an independent, nonprofit research firm specializing in long-term forecasting. A project called "New Consumer, New Genetics" seeks to give a thorough account of the implications of personalized nutrition on the food, pharmaceutical, consumer product, and health care industries. Their report *From Nutrigenomics Science to Personalized Nutrition: The Market in 2002* explores the likely impact of the emerging science of nutrigenomics on the consumption habits of Americans in the next decade. In *The Future of Nutrition: Consumers Engage with Science* the organization explores what course the science may take in the future, including research and development, intellectual property, food industry, consumer viewpoint, and consequences of the science and public demands.

genetic condition and to assess their health risks, (2) to provide support to individuals and their families, and (3) to facilitate informed decision-making in an effort to maintain a patient's sense of personal control (Wang et al. 2004).

Each of these dimensions of nutrigenomic service delivery is considered in this chapter. We explore a range of existing, emerging, and possible models of service delivery, and we examine benefits and challenges for each. Four possible models are considered: the consumer model, the health practitioner model, a pair of blended models, and the public health model. Each model has strengths and weaknesses, and each would be applicable to varying degrees of success, depending on a host of social context in which it is used. For example, where nutrigenomics services are not offered currently, one delivery model might appear to be better than the others, but whether that model is the best, or will continue to be used, remains to be seen. The discussion here steers away from conceiving of what would be the best service delivery model in any particular jurisdiction. Instead, we present the strengths and weaknesses of each potential model. We thus provide generalizable, and hence more widely applicable, results from our analysis of service delivery alternatives. Equally, more durable recommendations arise from our approach because we do not focus on short-term outcomes of specific instantiations of any of the models.

4.2 CONSIDERATIONS FOR NUTRIGENOMIC SERVICE DELIVERY

When considering the best model or models for delivering nutrigenomic services, we must understand the strengths and weaknesses of each model, including the strength of the science, the regulatory environment, the human resource capacity, the professional competency, the health care funding policy, professional politics, professional culture, and consumer attitudes and preferences. Together, these factors are relevant in the assessment of any particular model of service delivery, and they also furnish us with a foundation for making comparisons between models, which are summarized in Table 4.1.

4.2.1 Strength of the Science

In Chapter 2 we discussed in detail the strength of nutrigenomic science. In that discussion it was pointed out that within the community of nutrigenomic scientists, a number of concerns have been raised about the strength of nutrigenomic science and its downstream impact on nutrigenomic applications.

TABLE 4.1. Nutrigenomic Service Delivery Models

Service Delivery Models	Description	Benefits	Challenges
	Consumer Model		
	• Tests are most commonly sold over the Internet and by mail directly to consumers without the involvement of health professionals. • The self-administered test kits usually require consumers to collect a sample of their DNA by cheek swab. The sample is then mailed to the company for analysis.	• Tests promote the value of autonomy through the expansion of consumer choice. • Nutrigenomic tests are convenient and are easy to purchase and use. • Test results mailed home promote confidentiality and privacy.	• Private genetic testing may have negative consequences for public health care systems. • Tests may be offered inappropriately if the health benefits of such tests have not yet been established. • Tests lack standardization and the ability to check their credibility. • Consumers may have difficulty understanding the results and choosing health options without proper advice from health professionals.
	Health Practitioner Models		
Genetic specialists	• Genetic specialists are trained in medial genetics. • Specialists typically offer diagnostic or prognostic services and genetics education. Where needed, they offer psychosocial support in both pre- and post-test patient counseling.	• Genetics specialists are well positioned to offer nutrigenomic services because they already have training and expertise in risk assessment and psychological counseling.	• These is a lack of trained personnel. • Resources are inadequate. • Knowledge of nutrition, food choices, and menu development is lacking.

TABLE 4.1. (*Continued*)

Service Delivery Models	Description	Benefits	Challenges
Primary care practitioners	• These are the health professionals who traditionally have the primary responsibility for disease prevention, initial assessment of medical problems, and long-term care: including family physicians, nurses, pediatricians, and internists.	• Primary care providers are well positioned to provide dietary intervention advice, and typically have more experience in this field than do traditionally trained genetic specialists. • A family-centered approach to health care provides a valuable tool for risk assessment.	• There is low professional receptivity to genetic testing. • There is a lack of genetics and nutritional science education. • There is a lack of knowledge in nutrition, food choices, and menu development. • There is insufficient time for the proper delivery of services. • There is a lack of reimbursement for this type of service.
Nutrition specialist providers	• Specialist provides include physician nutrition specialists and registered dietitians.	• Nutrition specialists are trained to advise patients on food, diet, and nutrition information.	• There is a lack of genetics education and training in genetic counseling. • Personal counseling about diet–gene interactions is not currently reimbursable.
Blended Models			
Health provider team	• There is a division of responsibilities among various health care professionals, who function as a team.	• The system allows providers to strengthen services by taking advantage of different training and practice experiences.	• This is a complex model that may have costly and complex infrastructure requirements to be maintained.

| Consumer–practitioner services | • Health care teams often consist of physician, physician extenders, registered nurses, registered dietitians, and genetics specialists at tertiary medical centers.
• Nutrigenomic testing and tailored nutritional advice are offered directly to the consumer.
• Physicians or other health care providers are consulted after the consumer has taken a test, to discuss the results. | • The system is cost-effective, reduces the educational burden for any one profession, and distributes time demands among professionals.
• Since the average person lacks sufficient genetics education, a physician is required to advise on implications of tests results. | • Demands on physicians and other health care providers increase, because they are required to be genetic counselors and nutritionists.
• Any additional costs in this area may result in the neglect of other health needs. |

Public Health Model

| | • This model includes government agencies and public health practitioners.
• Nutritionists help develop policy, conduct research, and develop communications for the public. | • The model has the ability to make certain medical services widely available to the public. | • There may be trade-offs between privacy and the community good. Certain groups might face stigmatization.
• There is a reluctance to accept testing without evidence-based guidelines that address the relative benefit versus the cost or harm. |

Science quality is principally a problem for regulation, but it has an effect on the delivery of services. The quality of the science will play a central role in helping people to decide which tests, and what kind of advice, should be made available. As we mentioned in Chapter 2, the crucial issue is the strength and number of the nutrient–gene associations that would be used to substantiate health claims about dietary changes. We have suggested that as the associations become stronger and there is more empirical evidence that dietary interventions reduce disease incidence, nutrigenomics will move along the continuum from wellness testing to medical testing. This shift may require more sophisticated testing and counseling as the information requirements, many of which were considered in Chapter 3, take on new dimensions. It may also be the case that health care professionals will become more interested in nutrigenomics if tests of higher penetrance and higher predictive value become commonplace. On the other hand, if nutrient–gene associations that are highly predictive do not become the norm, another situation may emerge. Low-penetrance genes are much more prevalent in the general population, and pharmaceutical firms have predicted that most commercial gains and, consequently, most commercial interest, will be in test services for common low-penetrance genes which many people share (Hinsliff 2002).

4.2.2   Regulatory Environment

We have already alluded to the impact of regulations on service delivery. Indeed, the most obvious point at which nutrigenomics will be regulated is in upstream research, as discussed in Chapter 3, and in downstream service delivery. The latter, of course, is a key focus of regulators because the public is most exposed to the potential for harm when they receive nutrigenomic services. The strength of the science, again, has much bearing on the preferred service delivery model. Services offered direct to the public might be considered innocuous as long as the genes tested and the health advice given has a low potential for harm. For example, if a genetic test indicates that a specific supplement would be beneficial, this outcome is not likely to raise concern. If services are offered that lead to difficulties in interpreting genetic information, and dietary recommendations are made that have the potential for harmful side effects, greater regulatory scrutiny might drive nutrigenomics into a more heavily regulated context.

4.2.3   Human Resource Capacity and Professional Competence

Nutrigenomics lies at the intersection of several fields, including genetics, physiology, nutritional sciences, and dietetics. It is also a new field, which means that very few programs exist for professional training. For example, the University of California–Davis and the University of Stellenbosch have run nutrigenomics short courses at the graduate level, but only in the past few years. Even if there is a demand for nutrigenomic services, there may not be enough trained personnel. This is a serious issue, because the availability of skilled personnel will have a major influence on what types of services are available and to what degree. For example, proper provision of nutrigenomic services requires a complex set of skills and knowledge bases to cover such issues as pretest counseling; informed consent; test administration; test result interpretation and communication; posttest counseling, including both dietary and genetic counseling; and in some cases, long-term support. Nutrigenomic services will be delivered effectively and to capacity to the extent that there are trained people in the field, and they will not attract undue scrutiny of regulators as long as there is demonstrable professional competence.

4.2.4 Funding Policy

Interest in nutrigenomics has risen in the midst of what is sometimes described as an obesity and diabetes epidemic, which has severe epidemiologic and economic consequences (Yach et al. 2006). Public health officials grapple less with infectious diseases than they do with lifestyle diseases associated with poor diet and a lack of exercise. The reality of medicine is that it continues to treat disease and relieve symptoms, but this paradigm may be softened as preventive medicine finds its way into practice. A precondition of widespread adoption of preventive medicine is that public health authorities and insurers recognize the impact of prevention on health care financing. Admittedly, this could be double-edged. Any impetus toward a policy that favors preventive medicine may recoup monetary and health rewards, and if personal responsibility becomes a critical access condition, it may also change the face of how health care is funded, and to whom. Instances in which physicians have refused to treat patients who smoke because of concerns about the continuing effects of smoking on their response to interventions and future health (National Institute for Health and Clinical Excellence 2005) may become a model for the role of personal responsibility in decision-making regarding the allocation of scarce resources (CTV 2006). Nutrigenomic service delivery could certainly be affected by such changes in attitudes about health care and resource allocation, potentially in its favor if nutrigenomics' promise of prevention captures attention and is based on scientific results. Nutrigenomics may also propel health care funding decisions in the direction of prevention.

4.2.5   Professional Politics and Culture

Given that nutrigenomics draws on a number of professional competencies, it is foreseeable that there will be struggles over practice boundaries. For example, it is unclear whether general practitioners and genetic testing specialists ought, in general, to be responsible for administering genetic tests (Greendale and Pyeritz 2001). In the case of nutrigenomics, the advent of specially trained nutrition specialists or dietitians could complicate the question further. Equally, there could be some disagreement about whether genetic and nutrition counseling would best be delivered concurrently, and by what kind of professional. There may be a ready response to some of these questions if the issue has arisen from the regulation of, for example, other genetic tests and can be applied to nutrigenomics. Another factor is that regulations may change as professional competencies change. Professional perceptions of the acceptability and/or need for such tests will also play a role. At present, general practitioners appear to be cautious about the potential benefits of genetics in clinical settings, and are often not interested in becoming better educated in the field (Burke and Emery 2002). Other professionals, such as registered dietitians, may be more accustomed to thinking in terms of prevention and therefore more open to providing nutrigenomic tests. Levels of professional openness to this technology depend on underlying judgments about the value of disease prevention and health promotion and the role of genetics and nutrition in influencing illness and health. These judgments are colored in part by the professional culture, particularly during initial professional training.

4.2.6   Consumers and Patients

Nutrigenomics is often described as having two main points for service delivery: in the medical context, which would serve patients, and in market context, which would serve consumers. Educated, health-conscious, and affluent early adopters might drive the development of the nutrigenomics market, whereas public health officials and health insurance companies hoping to entrench preventive medicine might drive the medical model. Which mode of service delivery will become dominant is likely to be jurisdiction dependent, but it will certainly be shaped by a number of consumer and patient attitudes. Among these attitudes, one can anticipate important differences about the benefits and risks associated with nutrigenomics, people's level of trust in practitioners as compared to private companies operating in an open market, the level of interest in genetic determinants of health and prevention, and concerns about consent and confidentiality. Systematic research on public attitudes is only starting in a number of countries, so there is little empirical research on which to base claims about public attitudes toward nutrigenomics.

4.3   FOUR ALTERNATIVE MODELS

Four alternative models of nutrigenomics service delivery will be considered here: the consumer model, the health practitioner model, the blended model, and the public health model. Each model is discussed below as if it were an archetype of service delivery, a way of representing the case that is blind to many of the contextual details that would be salient to the success of one service delivery model over another. The merits of this approach are that it is generalizable and permits cross-model comparisons for strategic thinking about the advantages and pitfalls of each service delivery model. The contrasts we draw, however, are not to suggest that only one approach to service delivery will reign. We suspect, instead, that nutrigenomics will deliver different models simultaneously, depending on the particular service, regulations, and interests of the consumer or client base. Some tension between service providers may arise as nutrigenomics service delivery evolves, and we have anticipated this in our discussion of the models.

4.3.1   Consumer Model

In Chapter 2 it was pointed out that the general concern within the nutrigenomics scientific community was that it is too soon to be delivering nutrigenomic tests to consumers. Nevertheless, nutrigenomics is drawing considerable media attention. Nutrigenomics has been discussed in Oprah Winfrey's magazine *O* (Kapp 2005), popular science journals such as *Wired Magazine* (Duncan 2002), *The Scientist* (Pray 2005), and *Technology Review* (Kummer 2005), and in newspapers (Rundle 2005). Academic conferences on the scientific and ethical issues involved in nutrigenomics are also receiving media attention (Check, 2003; Romain 2003). As the public has access to more information about nutrigenomics, they may seek out nutrigenomic services. Finding little awareness or support from most health care practitioners, early adopters often turn to the Internet, where they will find a handful of companies offering direct-to-consumer nutrigenomic services. Some have speculated that in what perhaps might be described as the last step in a chain of personalized nutrition, knowledgeable consumers will not only avail themselves of nutrigenomics but may also seek novel and improved products from primary producers to get higher food functionality (German et al. 2004).

Sciona Profile

Sciona Ltd. began as a small biotechnology company in England on the day the rough draft of the human genome was announced in 2000. The company founders, Rosalynn Gill-Garrison, Keith Grimaldi, and Chris Martin, believed that the human genome project heralded the future of human biological sciences and medicine. The opportunity seemed ripe to start a company whose mandate was to translate genomics into useable technologies that would benefit the public. Together with a half dozen employees, Sciona embarked on what appeared to be a human genomics and genetics gold rush.

The company's business strategy was to combine a dietary assessment and lifestyle questionnaire with a genetic test. The genetic test consisted of a cheek swab, which was returned to Sciona with the questionnaire, which was then analyzed for roughly a dozen polymorphisms. The results were returned to the person by mail with company contact information. Offered first through pharmacies such as Boots and the Body Shop in the United Kingdom, this direct-to-consumer test attracted the attention of the advocacy group, Gene Watch, and gained some notoriety in the U.K. daily, The Guardian, in 2002. The Human Genetic Commission, in a 2003 review of direct-to-consumer genetic tests, recommended that tests similar to Sciona's should be regulated to prevent situations marred by incomplete consent being collected or misleading advice being given.

Recognizing that it was caught in a regulatory vacuum that allowed concerns to persist, Sciona entered into a discussion with the Human Genetics Commission, but was not able to engage GeneWatch in a dialogue about the issues raised in the public domain. Sciona moved away from direct-to-consumer marketing, and started to offer its services through physicians. The company continued to raise money, conduct research, and entered into new research partnerships in the United Kingdom until it moved to Boulder, Colorado in 2004.

The company continues its research and development initiatives with international partners, and has offered a new test. A commercial retail currently offers five different nutrigenetic tests that test for a total of 24 variants in 19 genes. Most of these tests are offered direct-to-consumer, but Sciona also offers them through integrated pharmacies and through physicians depending on the country in which the test is being sold. Sciona provides the test in partnership with a private laboratory, which does the genetic test. Sciona provides the analysis of the genetic test results, in light of the information reported in the lifestyle and dietary questionnaire by the client using a patent-protected computer algorithm.

Aside from concerns by the nutrigenomic scientific community regarding the readiness of the science for applications, there is also considerable controversy about use of the Internet as a platform for advertising and delivering services. Most direct-to-consumer or do-it-yourself testing kits are sold over the Internet and by mail, often without the involvement of health professionals. The self-administered test kits typically require consumers to collect a sample of their DNA with a cheek swab. The sample is mailed back to the company for analysis, and consumers receive an interpretation of their test results. These are not commonplace, however. In a recent survey of direct-to-consumer genetic tests on the Internet, Gollust (2003) found that of the 105 sites found, only 14 offered tests that were explicitly linked to health. Nevertheless, some consider and because direct-to-consumer delivery of genetic testing services irresponsible because it is premature and because what is offered is not necessarily delivered (Vineis and Christiani 2004). Another criticism is that state support for direct-to-consumer services is a poltical act in which the state deliberately substitutes individual responsibility for reform and repair of publicly provided health care.

Benefits of the Consumer Model

The chief argument in favor of the direct-to-consumer model of service delivery is that it is a ready means of access to health information and technologies that could be beneficial. At root, the idea is that if one takes the autonomy of individuals seriously, one has to recognize their right to determine what should be done to their bodies (Stevenson 1999). Nutrigenomic tests marketed directly to the public is one way of transmitting knowledge quickly from science and technology research and development into the public domain. Access to this information arguably promotes individual autonomy, and autonomy is enhanced when people have a greater range of options over which to exercise choice. In addition to the autonomy argument, it is often argued that access to information in a nonthreatening and end user–controlled environment has an additional benefit—it empowers individuals to make healthy choices, to confront their eating habits with greater honesty (Kearney and McElhone 1999), and to make behavioral changes more easily (Koelen and Lindström 2005).

Some have argued that access to one's genetic information is just as much a right as having access to other medical personal medical information. Buchanan observes that some have "endorsed a presumption that competent adults have the right to information about their genotype, if the information is accurate and its significance is conveyed to them in an appropriate manner" (Buchanan et al. 2002). To answer the quality assurance question that is being raised, it has been suggested that some consumers would feel more confident about the accuracy, timeliness, and relevance of the genetic information they pay for if there were a consumer charter for genomic services. As Ledley suggests charter-supported services could have far-reaching effects:

> Inherent barriers in the current healthcare system could inhibit the application of genomics to personal health. A radical reassessment of healthcare delivery models is necessary to ensure that the health benefits of genomics are realized by the greatest number of individuals in the shortest possible time. A model that empowers consumers with confidential knowledge of their own genome and the ability to make informed decisions concerning their health care represents the ultimate expression of a truly personalized medicine. It is also an effective strategy for circumventing the inadequate infrastructure for genomic services, improving the quality of care, and strengthening protections for individual privacy and autonomy. (Ledley 2002)

Even with a charter, the success of direct-to-consumer delivery of nutrigenomics will depend on careful tailoring of messages about nutrition (Peltola 2002). For direct-to-consumer to work, particularly when provide online, it will be critical to communicate with the public in ways that build trust, provide accurate information, and gain social acceptance (Bouwman et al. 2005).

Although enhancement of consumer autonomy and consumer empowerment are the main benefit of direct-to-consumer nutrigenomic services, other benefits can be anticipated. Direct access may contribute to the demystification of medicine and genetics for the layperson who takes the time and effort to learn, and may also encourage self-care (Janata 2003). One possible outcome is that direct public access to genetic tests could drive the public toward personal responsibility and partnership in health decisions with health practitioners (Human Genetics Commission 2003). Preventive health measures might be promoted in the process. As the Institute for the Future reports (2002), people who seek private testing services will already be highly motivated to make and sustain dietary lifestyle changes, and they will have the greatest likelihood of getting any benefits. Finally, the availability of direct-to-consumer testing makes it possible for people to take the test without having the results become part of their medical records. This speaks to concerns about privacy and confidentiality and about the possible risk of genetic discrimination or stigmatization discussed in Chapter 3.

Challenges for the Consumer Model

Concerns that nutrigenomic services ought not to be provided directly to consumers are generally motivated by worries that commercialization will bring tests to market prior to rigorous scrutiny of the test itself, the nutrient–gene association being assayed, and the health claims that are made about a person's test results and his or her lifestyle and diet. In many cases there is little risk of serious physical harm. If consumers and patients do not clearly benefit, or if they are mislead into thinking they have benefited when they have not, direct-to-consumer nutrigenomic services might be premature. This line of reasoning puts nutrigenomics in a catch-22 situation, since consumer use is one way of validating nutrigenomics. The catch-22 could be resolved if consumer behavior and health outcomes were monitored well after the test, but follow-up with consumers is not the norm.

Marketing and advertising directly to the consumer might, by virtue of the context, make the test to be consumed different from tests offered in a clinical environment. It may be that the marketing of tests, if realistic about what they offer, might play up the positive aspects of the test without proportionate consideration of the negative aspects. Concerns have been raised, for example, about test marketing that downplays the uncertainties associated with gene testing. Analysis of 28 leaflets about cystic fibrosis carrier testing from commercial and noncommercial organizations in the United States and the United Kingdom revealed that information from commercial organizations was more likely to lead to people taking the test than was information from noncommercial sources (Marteau 1999). Considerations such as these led the U.S. Task Force on Genetic Testing to recommend that advertising and marketing of genetic tests directly to the public be discouraged (Caulfield et al. 2001). Finally, there is the potential for financial conflicts of interest if the results of a test tend to lead to recommendations that products sold by the same vendor would be appropriate interventions. For example, a company may sell some of the nutritional supplements that it recommends as part of the dietary counseling that accompanies their analysis of genetic tests (GeneWatch 2003). The UK Human Genetics Commission warned that government agencies need to protect consumers from any abuses in such situations.

Formed in 1998, GeneWatch UK (www.genewatch.org) is a not-for-profit organization that monitors genetic technology development to ensure the safety of the public, environment, and animals. This institution has been very critical of the sale of genetic tests by scientific companies in the United Kingdom. In a January 2003 article in *The Observer*, GeneWatch was quoted as saying that certain genomic profiles were based on poor and imprecise science and were offered by a company that could stand to gain financially from recommending dietary adjustments.

Direct-to-consumer tests raise questions about how information will be interpreted by a consumer who may have no third-party assistance. It is obvious that direct-to-consumer nutrigenomic information should be accurate, understandable, and comprehensive. Studies of the interpretation the advertising and results of other types of genetic tests do not raise confidence levels. A number of potential problems, including misinformation about genetics, exaggeration of risks, the suggestion of an overly deterministic relationship between genes and disease, and the reinforcement of associations between certain diseases and ethnic groups have been identified (Gollust et al. 2002). These worries have led some professional groups to recommend that direct-to-consumer genetic tests not be encouraged (American College of Medical Genetics 2004). In addition, the potential for misinformation to arise is heightened by the current lack of genetic literacy on the part of the public, which may lead to conceptions of nutrigenomics that are based in the language of genetic determinism rather than that of long-term susceptibility (McCabe and McCabe 2004). Protection of consumers against false or misleading claims, it has been suggested, would require premarket approval for testing services, which is currently not required.

Finally, it may be the case that at the heart of direct-to-consumer provision of genetic testing and test-based services lies some conflict regarding the role of consumer vis-à-vis that of patient. This is a complex problem, for on the one hand, nutrigenomics can appear as innocuous as having a free blood pressure test at the pharmacy counter in a supermarket. The information learned may have very little impact on a person's behavior, and a blasé attitude may be less of an indication of indifference than a reflection of people's beliefs that medicine belongs in a clinical context and nutrigenomics in another, but certainly nonmedical context. If direct-to-consumer contact encourages people to think about their health issues in consumer rather than patient or clinical context, it could lead to new and perhaps unfruitful conceptions of health and well-being. It is worth considering whether it is problematic to think about health like a consumer if people begin to relate to their bodies as broken machines for which the outcome of treatment is repair, not health (Hudak et al. 2003). Because the consumer model promotes autonomy and empowerment, genetic counseling is less likely to occur within this model. Genetic counseling can be defined as a therapeutic and dynamic relationship between provider and client with the goal of facilitating the "client's ability to use genetic information in a personally meaningful way that minimizes psychological distress and increases personal control" (Wang et al. 2004). Without appropriate genetic counseling, consumers under this model are unable to make informed decisions about their health and therefore run the risk of not being able to interpret the results of a test and the health risks of the genetic condition in question.

4.3.2 Health Practitioner Model

One response to the challenges of direct-to-consumer testing is the introduction of appropriate regulatory safeguards in the marketplace. These may be difficult to develop for a rapidly changing field (as we discuss in Chapter 5), and they may even be more difficult to enforce. Another option is to abandon direct-to-consumer regulation by prohibiting the activity, or to displace direct-to-consumer activities, or in a third scenario, to move the most problematic aspects of nutrigenomic services under the purview of an appropriate health care provider. The delivery of genetic test services through a professional "gatekeeper" does not mean that no private interests are involved but that the point of access for nutrigenomic services are controlled by professionals who are already regulated.

Health care practitioners with different training (e.g., genetic specialists, primary care practitioners, and nutrition specialists) might each claim to have a legitimate claim on being the profession most suited to delivering nutrigenomic services. All can share the view that nutrigenomics is another tool that can assist health professionals in helping people mitigate or prevent disease, and achieve optimal health and well-being.

Genetic Specialists

Before the "genomics revolution," genetic services were delivered primarily by specialized clinical geneticists, genetic counselors, and genetic nurses. Clinical geneticists are physicians trained as medical geneticists or who have doctorates in genetics. They typically offer diagnostic or prognostic services. Genetic counselors provide genetics education and, where needed, psychosocial support in both pre- and posttest patient counseling (Biesecker and Marteau 1999). Genetic counseling services are provided by a number of professionals, depending on the jurisdiction. In Canada and the United States, most counseling is done by genetic counselors trained at the master's level. In Australia and the United Kingdom, counseling is usually provided by nurses working with medical geneticists. In many parts of Europe, genetic nurses or social workers act as genetic counselors (Biesecker and Marteau 1999).

BENEFITS OF DELIVERING TESTS THROUGH GENETIC SPECIALISTS. In the wake of the Human Genome Project, the traditional focus of clinical genetics has forecast expanding from the diagnosis and prediction of rare, often untreatable conditions, to determining susceptibility to common, often preventable conditions. Nutrigenomic testing falls into the latter category.

Genetics specialists are well positioned to offer nutrigenomic services because they already have the requisite training and expertise in risk assessment and psychological counseling. Although historically, genetic diagnosis has been based on a physical examination or a family history, molecular testing is increasingly being used by specialists for this purpose (Biesecker and Marteau 1999). Given that genetic tests are already a familiar tool for this group of specialists, incorporating nutrigenomic testing technologies into their practice should not be difficult.

CHALLENGES OF DELIVERING TESTS THROUGH GENETIC SPECIALISTS. Three challenges face the specialist model of service delivery: a lack of trained personnel, inadequate resources, and insufficient familiarity with nutritional science. Until recently, clinical genetics focused almost exclusively on single-gene disorders and chromosomal abnormalities. The prevalence of these conditions is sufficiently low in the general population that most people have little concern about developing them. Consequently, genetic specialists in countries such as Canada and the United States, although small in number, have been able to meet the genetic health care needs of the entire population (Guttmacher et al. 2001). Nutrigenomic tests, by contrast, look for diet–gene interactions linked to wellness promotion and disease susceptibility. Furthermore, if and when the research warrants it, it may be possible to offer genetic testing for susceptibility to common diseases, with complex causes, such as a variety of cancers and heart disease. These are conditions so prevalent that almost everyone has personal concerns about developing them. This expected increase in the demand for tests has led some within the profession to recognize the need for new methods of service delivery.

Without significant changes to existing funding structures, the specialist model is unlikely to become an established method of service delivery. In most countries that have a genetics specialization, funding is not provided for the training of genetic counselors and there is no reimbursement for the provision of such services. This funding situation contributes to the low number of residents in medical genetics. In 2001 there were only 45 medical genetics residencies in the United States, less than half the number of a decade before. During the same period, there were 2000 genetic counselors in practice but only 160 counselors and 50 clinical geneticists in training per year (Guttmacher et al. 2001). This extremely low number of specially trained professionals is expected to contribute to a human resource shortfall that some predict will reach crisis proportions when combined with the expected increase in numbers of test users. The almost universal lack of reimbursement for genetic counselors is also a disincentive for future practitioners, a problem magnified by the complex and time-intensive nature of genetic counseling.

Clinical genetic services and accompanying genetic counseling are currently offered primarily for the prediction and/or diagnosis of rare and generally untreatable conditions (Biesecker and Marteau 1999). Special training is needed for such genetic counselors because test taking often reveals grave consequences for patients. By contrast, nutrigenomic tests aim to identify increased disease susceptibility for the purpose of providing advice on lifestyle changes that can reduce disease risk. Consequently, test taking is unlikely to be an agonizing decision for the patient, potential outcomes are not likely to be life threatening, and nutritional interventions are often available. Because of these differences in risk profile and the availability of effective interventions, it is not clear that the administration of nutrigenomic tests needs to fall under the purview of a specially trained genetic counselor, except where susceptibility to a serious condition is involved or in cases where testing for a polymorphism may raise implications for multiple diseases.

Primary Care Practitioners

In light of the challenges facing genetic specialists, in particular the anticipated increase in consumer demand for genetic services, many health professionals are predicting a general shift in the provisions of genetic tests away from a specialist to a primary care model of service delivery. It is conceivable that this shift will encompass nutrigenomic testing. Primary care providers are those health professionals who have the main responsibility for disease prevention, initial assessment of medical problems, and long-term care. These health professionals include family physicians, nurses, pediatricians, and internists.

BENEFITS OF DELIVERING TESTS THROUGH PRIMARY CARE PRACTITIONERS. The clinical value of nutrigenomics will arise from dietary interventions which improve patients' health over the long term and lessen susceptibility to disease. Primary care providers are already accustomed to advising patients to engage in healthier behaviors and typically have more experience doing this than do traditionally trained genetic specialists (Guttmacher et al. 2001). If only for this reason alone, primary health care practitioners are somewhat better positioned than genetic specialists to deliver nutrigenomic services.

Primary care providers who make use of genetic susceptibility tests would be able to integrate this new knowledge into their practice as part of the longer-term relationships with patients (Nature Medicine 2003). Assuming that there is continuity of care between a patient and a primary health care practitioner over a long period, there is the opportunity to evaluate the clinical impact of genetic information and the impact that this information has on people's capacity to make lasting dietary changes. In many cases, primary care providers are also accustomed to using a family-centered approach to health care. They generally know both the patient's and the family's history, so this would allow them to tailor genetic information to the particular needs of patients (Guttmacher et al. 2001). Finally, the practice of follow-up visits to discuss results and monitor progress is ingrained in most primary care practices, and this approach will be needed to deliver nutrigenomic services properly, including dietary and lifestyle advice. For a number of reasons, it seems reasonable to expect that if well supported by the science, nutrigenomic testing could eventually come to be viewed by primary health practitioners much like any other medical test for risk prediction. If the anticipated shift toward preventive medicine gains a foothold, nutritional genomics may find a comfortable home in the offices of primary health care practitioners.

CHALLENGES OF DELIVERING TESTS THROUGH PRIMARY CARE PRACTITIONERS. Despite the advantages describe above for having nutritional genomics offered through primary health care practitioners, some serious professional competency obstacles may stand in the way of easy uptake of nutrigenomics into these professions. These challenges include low professional receptivity to genetic testing among primary health care practitioners, their lack of genetics and nutritional science education (Korf 2002), and the lack of time required for the proper delivery of services and potential lack of private or public health system reimbursement for nutrigenomic services.

As nutrigenomic tests become more widely available, greater numbers of primary care practitioners will find themselves in unfamiliar territory. Nutrigenomics offers presymptomatic testing that will not provide diagnoses of current diseases but will give estimates of a patient's propensity to develop certain diseases based on a selected set of genetic variants. Perhaps this oversimplifies the problem. Very few tests are for monogenic disorders with deterministic outcomes, such as Huntington's disease. There are a few tests for high-penetrance variants, such as the BRCA genes associated with breast and ovarian cancer. The crucial difference between the two is that BRCA positive tests do not lead to deterministic predictions about future disease, but instead, indicate that disease is highly probable.

Nutrigenomic tests such as the BRCA test are less strongly predictive, but what nutrigenomic and BRCA tests have in common is that neither is diagnostic for the presence of disease. A positive result in either case indicates a propensity for disease, not the presence of disease. The difference, however, is that primary health care practitioners will be more inclined to order tests for high-penetrance genes if they believe that it will support a diagnosis or play a role in disease prevention through, for example, prophylactic surgery. If physicians are by and large not used to ordering genetic tests, even in cases of highly penetrant genes, they will be even less likely to order tests for long-term disease susceptibility. Since there may very well be no disease symptoms, or other underlying but detectable causes of disease, a physician prescribing tests in response to patient demand may be uncomfortably unable to distinguish between the sick and the worried well (Castle 2003).

Lack of appropriate education in genetics and nutrition has been identified as one of the biggest challenges for this service model. Practitioners need to know how to understand and communicate susceptibility information and how to use this information in combination with nutrition information to advise their patients about necessary changes in diet. This problem is twofold. First, most general health practitioners lack expertise in clinical genetics and molecular testing and have a poor grasp of basic genetics, probability, and risk (Murray and Botkin 1995). Service providers need to understand and communicate complex probabilistic information, including how to quantify risk, how to frame test outcomes, and how to deal with false negative and false positive test results (Marteau 1999). Because understanding and interpreting predictive tests requires a fairly sophisticated understanding of genetics, these tests are highly susceptible to misinterpretation, even by medical professionals. One study that examined the use of a commercial genetic test for a mutation in a colon cancer gene found that physicians incorrectly assessed the test results in 31.6 percent of cases studied (Guttmacher et al. 2001). This insufficient knowledge about genetic tests administered by doctors has been documented in a number of clinical settings, including obstetrics and oncology. When asked, primary care practitioners identify lack of genetics education as one of the greatest deficits in their professional training.

Second, primary health care practitioners often lack sufficient nutritional science knowledge that would be required to interpret a nutrigenomic test. This is particularly true of physicians, who often lack confidence in their knowledge and are reluctant to give nutritional advice to patients (Kolasa 2005). In one study, American physicians on average spent less than a minute discussing nutrition with their patients (Flocke and Stange 2004). In another study, the time spent in nutrition counseling in primary care settings during an average office visit lasting 10 to 16 minutes is usually less than 5 minutes per patient, the average time being 1 minute (Eaton et al. 2003). A recent in-depth analysis of the role of nutrition education in medical schools and postgraduate training programs showed a large deficit in practical nutrition knowledge (Walker 2003). This gap in training persists despite the fact that there is evidence that the public thinks that medical professionals have

greater knowledge of nutrition and place it high among the factors important to the maintenance of health and the management of disease (Halstead
1999).

Strikingly, one international conference on nutrition and medical practice
in the late 1990s found that physicians from 11 countries with well-devel-
oped medical institutions were only vaguely aware of a link between proper
diet and health. For better or worse, another study of nutrition interventions
by primary care practitioners in the United States concluded that levels of
basic nutritional knowledge were fairly good (Moore and Adamson 2002).
In addition to there being variable or low levels of knowledge about nutrition, there are other barriers that prevent primary care practitioners from
providing dietary counseling: lack of time, lack of patient compliance, inadequate teaching materials, lack of nutritional knowledge, and insufficient
reimbursement (Kushner 1995).

Professional attitudes about nutrition as a means of disease prevention also
stand as a barrier to nutrigenomic testing. Nutritional advice that accompanies most nutrigenomic testing is for primary prevention. Primary care physicians assign this type of counseling low priority, except for pregnant women,
infants, and the very old (Truswell 1999). Family doctors participating in a
series of international conferences held during the last decade concluded that
secondary and tertiary prevention should be the focus for nutrition advice in
general practice. The assessment of patient risk and the provision of appropriate patient care are time intensive, and primary care providers do not have
time available for such services. The total time that physicians spend talking
with patients about all health matters is usually less than the time that genetic
counselors spend informing people about genetic tests and their implications
(Holtzman and Watson 1997). The pressure in many primary care settings to
see a high volume of patients does not lend itself to the kind of educational
and, in some cases, psychological supports that need to be provided for users
of nutrigenomic tests. The intense time demands of nutrigenomic test
services is also likely to be an economic disincentive for meeting an
appropriate standard of care.

Government and private insurance programs do not provide adequate
funds for health measures aimed at prevention. Almost no payer in the United
States reimburses for genetic counseling services, even when the service is
provided by a physician (Greendale and Pyeritz 2001). Coverage for dietary
nutrition services is not provided in many jurisdictions, although this may
be slowly changing with the 2000 addition of medical nutrition therapy as a
covered benefit for Medicare for specific medical conditions and the addition of registered dietitians as providers. Recently, Blue Cross/Blue Shield
of North Carolina announced their intention to cover nutrition services
(Kushner 1995). This lack of reimbursement is both a disincentive to the

offering of nutrigenomic tests and threatens to lead to substandard care in cases where the tests are administered. Insufficient reimbursement has been identified as one of the dangers of health practitioner–delivered services because it leads to such problems as inadequate pre- and postgenetic test counseling.

Nutrition Specialist Providers

When we refer to nutrition specialists, we include physician nutrition specialists as well as registered dietitians. The latter group are likely to be a significant source of nutrigenomics service delivery, particularly in the North American context, as the American Association of Dietitians and the Canadian Association of Dietitians grow steadily more interested in nutritional genomics. Dietitians are not unlike primary health care practitioners in the sense that they require continuing education credits to keep registered. Nutrigenomic training initiatives are already being developed that target dietitians, and this training will enable them to become the "clinicians of health" that some see emerging in dietetics practice in the future (German 2005).

BENEFITS OF DELIVERING TESTS THROUGH NUTRITION SPECIALISTS. Nutrigenomics would appear to have a natural home with nutritionists and dietitians. Effective dietary interventions combine nutrition education with behavior-oriented counseling, reinforcement, and follow-up to help patients acquire the skills, motivation, and support needed to alter their daily eating patterns and food preparation practices (U.S. Preventive Services Task Force 2003). The physician nutrition specialist has the requisite training in diagnosing nutrition disorders and nutrition and dietary counseling to effectively bring about healthy changes in eating patterns. Registered dietitians have a broad-based nutritional knowledge and are already accustomed to providing individual dietary counseling and to engaging in mass communication efforts to promote healthy diets. There have also been calls within the nutrition community to add the new genomics technologies to the tools of the trade as a way of further tailoring dietary advice to individuals (Goldberg 2000). Not only will this revolutionize what nutritionists and dietitians do but will have a corresponding impact on the field of nutritional genomics. As DeBusk et al. (2005) comment: "[T]he dietetics profession has an exciting opportunity that, if seized and properly executed, could enhance the scientific foundation of clinical practice, improve therapeutic outcomes and significantly expand career and economic opportunities for practitioners. The future of dietetics is unquestionably intertwined with nutritional genomics." To reach this level of professional competency in nutrigenomics, dietitians will have to embrace new educational opportunities, which some have likened to the advent of nurse practitioners (Skipper 2004).

CHALLENGES OF DELIVERING TESTS THROUGH NUTRITION SPECIALISTS. To be able to provide nutrigenomics advice, dieticians and nutritionists will need to be able to tailor dietary advice according to differences in genotype. This requires an understanding of genetic risk, gene–environment interactions, and the awareness of some of the ethical, legal, and social implications involved in genetic testing, and the ability to communicate all of this to patients (Burton 2003). Advances in genetics are providing the knowledge and tools needed to understand the relationships among nutrient intake, metabolism, genetic risk, and human behavior. Effective use of these resources will require the integration of interdisciplinary education for nutritionists and dieticians to produce specialists called *nutrigenomicists* (Arab 2004).

Developing competency in nutrigenomics is the first step toward securing reimbursement for counseling for nutrigenomics, counseling that currently is not reimbursed (Patterson et al. 1999). According to Hilary Burton of Cambridge University's Public Health Genetics Unit, however, genetics is generally not seen as a priority for dietitians and nutritionists, due to the complex nature of the subject and the limited resources available to develop educational programs (Burton 2003). A recent study concluded that there is next to no genetics training in nutrition education, let alone training in nutrigenomics. This raises a question about whether it is time to train dietitians in genetics early in their professional instruction (Jenkins et al. 2001; Vickery and Cotugna 2005). According to the Society for Nutrition Education, a major concern is that nutrigenomics may create the false impression that dietary guidelines are irrelevant except when their use is reinforced by a genetic test. It is therefore critical that as personalized nutrition becomes more of a reality, it is integrated into science-based dietary guidelines.

4.3.3 Blended Models

Blended Model 1: Health Provider Team

In light of the many tasks involved in nutrigenomic service delivery, it may be impractical to expect that any single health practitioner can have the requisite knowledge base and time to provide the full range and level of care required to ensure safety and efficacy. Instead, a carefully crafted division of labor among dissimilar health care professionals functioning as a team may be a more appropriate method of delivery. The chief benefit of this kind of team approach would be that it would allow providers to strengthen services by taking advantage of different training and practice experiences, knowledge bases, talents, emphases, and perspectives on genetics, nutrition, and their intersection in nutritional genomics.

Such a team-based approach to genetic care could operate using a simple referral system or a more actively integrated approach. For nutrigenomic tests used to determine disease susceptibility where the association is strong and/or the disease is traumatizing, the team might include the primary care physician, an expert in the disease for which the risk is being assessed, a nutrition expert, and a genetic counselor. Team members would need to set criteria for determining when primary care practitioners should draw on the expertise of other members of the genetic care team. In the United Kingdom, for example, family physicians are currently able to delegate nutrition counseling to nurses who work on the same premises (Truswell 1999). In the United States, physicians often refer to registered dietitians who work within the same health care institution. This is one way of designing a system that is more cost-effective, reduces the educational burden for any single profession, and spreads the time demands among professionals. Of course, this is also the main drawback to the blended health provider team—it requires institutional infrastructure to maximize the flow of communication between specialists and to the patient. This infrastructure may be costly to set up and maintain unless it sustains special fields other than nutrigenomics.

Blended Model 2: Consumer-Practitioner Services

In a consumer–practitioner service delivery model, nutrigenomic testing and tailored nutritional advice is offered directly to the consumer. Primary health care practitioners would then consult after the consumer has taken the test to discuss the results and diet recommendations, sometimes at the suggestion of the test company and sometimes on the consumer's initiative. The benefit of this approach is that it puts a patient into the hands of a professional primary health care practitioner who will can provide referral and follow-up consultations to the outsourced nutrigenomic testing. The drawback of this approach is that it puts pressure on primary health care practitioners to become capable interpreters of genetic tests and providers of nutrition information. Neither field, as we discuss elsewhere, is historically a strength of most primary health care practitioners.

Benefits of Blended Models

The average person lacks sufficient genetics education and, in particular, lacks adequate understanding of the predictive nature of genetic tests. This

Gene Care Ltd.

Gene Care was founded in Cape Town, South Africa in 2001 by Marithe Kotze. Kotze was a professor of human genetics and head of the department at the University of Stellenbosch, but resigned in 2002 to devote her energies full time to Gene Care. The company's staff of approximately half a dozen full time employees are engaged in research and development through partnerships and contracted research with universities and private companies. In the latter case, for example, Gene Care has developed a proprietary genetic strip assay with ViennaLab.

The company's principle efforts are focused on pre-symptomatic genetic testing for common polymorphisms associated with disease progression. Gene Care offers a wide variety of genetic testing services, including tests for paternity, genealogy, and for ethnicity-specific disease-genes, such as Tay-Sachs. In this respect, Gene Care's operates as a private genetic testing company capable of adapting its laboratory to the needs of its clients or research partners.

The nutrigenomics focus of the company is principally concerned with cardiovascular disease. Gene Care offers a test for several common polymorphisms in a number of genes, including APOE4, which are associated with elevated risk when coupled with poor dietary and lifestyle choices. Gene Care's strategy is to test for genetic variants on a semicustomized basis by offering tests for up to 12 variants in 10 genes, and the genes tested are contingent on family context, clinical information such as cholesterol levels, or lifestyle factors. The genetic test results are incorporated into a risk assessment and management strategy for cardiovascular disease already in use by physicians and dietitians. Gene Care offers the results and partial interpretation of the data to health care professionals using their services, but are not themselves in direct contact with the patient.

Gene Care offers its nutrigenomics testing services through HealthNet, a South African network of physicians and dietitians. HealthNet's objective is to combine genetic testing services with lifestyle intervention to prevent or mitigate the harmful effects of polymorphisms associated with elevated risk for heart disease when combined with poor diet and lifestyle choices. Gene Care provides the genetic testing services on a referral basis, meaning that a consulting physician and possibly also a dietitian would be involved in pre- and post-test patient counseling and care. The contracted services are paid for by patient's private health insurance, and the program has been supported by NetCare, a major integrated health care organization in South Africa.

means that the average consumer who accesses direct-to-consumer nutrigenomic tests may have a difficult time fully understanding the test results and their implications. For example, in cases where the person has a genetic mutation but is asymptomatic for any given disease or condition, a health practitioner may be needed to help to place the test results in their proper context. For this to work, the same autonomous decision-making, initiative, and empowerment that brought the consumer to the test in the first place is being relied upon to carry the consumer through to seeking assistance and then accepting posttest counseling. This would represent a very high degree of success for personalized nutrition and medicine were it to happen. As Marteau et al. (2004) report, the impact of genetic test results, even for those aware of their underlying condition, is highly motivational, even though it might predispose people to look for a medical solution rather than change their lifestyle.

Challenges for Blended Models

Consumers who seek services outside the clinical context will perhaps pose the greatest challenge to primary health care practitioners. Suppose that a user of a direct-to-consumer nutrigenomic test finds that he or she has a mutation for which dietary intervention is recommended but that require further consultation with a physician will be required. The consumer will then switch modes to being a patient, but the burden of correctly interpreting nutrigenomic advice generated outside the clinical context suddenly falls on the physician. Without the consultation, a patient may make inappropriate decisions with potentially grave repercussions. Conversely, successful consultation depends on the physician's ability to interpret properly genetic tests that they did not order and about which they may know little or nothing. If it is unreasonable to ask physicians to double as proficient genetic counselors and to gain detailed knowledge in the more narrowly focused field of nutrigenomics, will patients learn to trust sources other than their physicians? As it is, early adopters of nutrigenomics tend to have existing health and nutritional knowledge collected from sources other than their physician.

A blended model of service delivery raises resource concerns as well. In public health care settings, patients who pay for and take nutrigenomic tests privately, then approach health practitioners for additional advice and interventions, will impose further costs on that health care system. The existence of a private market reduces the economic efficiency of the public system to the extent that private companies are allowed to profit from this supplementary use of publicly funded health services. In a climate where resources are scarce, these additional costs may result in the neglect of other basic health needs with which they compete.

4.3.4 Public Health Model

Major public health advances were made in the twentieth century through simple initiatives such as purer drinking water, sanitation, and immunizations, resulting in a doubling of the average life expectancy (Hall 2003). During the same period, a number of noncommunicable diseases, led by cardiovascular disease and cancer, supplanted communicable diseases as the foremost health problem in industrial nations. In light of this shift, public health professionals have increasingly turned their attention to environmental determinants of health and disease, such as diet. In doing so, public health is carrying out its historical concern of prevention by dealing with environmental causes (Breslow et al. 1995). Government agencies and public health practitioners have begun to recognize the potential benefits of genetic technologies, and a new area called *public health genetics* is beginning to have influence in the United Kingdom and the United States. Public health genetics is about applying genetics and molecular biotechnology to improve the public's health and aid in disease prevention, and is a common public health initiative.

Benefits of the Public Health Model

Public health measures include making certain medical services widely available to the public and encouraging health-promoting behavior (Breslow et al. 1995). Nutrigenomics screening would support both measures and could be offered as part of a population-based prevention program. Theoretically, nutrigenomic tests could play a role in the prevention of disease by offering entire populations or targeted subgroups assessments of their susceptibilities to certain common diseases, combined with dietary advice.

Genetic screening uses the same assays as those employed for genetic testing, but it has a different target population. Screening is normally directed at groups of people with latent, early, or asymptomatic illness. Precedents exist for this type of public screening for genes affected by diet. Screening for PKU, a rare metabolic disease that causes mental retardation unless promptly controlled by a special diet, has been part of public health programs in many nations since the 1960s. Other examples include screening at-risk women for genetically related breast cancers, hemochromatosis screening in adults, and preconception or prenatal carrier screening for Tay–Sachs, cystic fibrosis, and sickle cell disease. Nutrigenomics screening, by contrast, would probably involve voluntary screening of healthy adults for susceptibility to late-onset diseases. Francis Collins, director of the National Center for Human Genome Research in the United States, recently predicted that "in the near future physical examinations for 18-year-olds will include DNA tests for diseases with a genetic component and that physicians, in the interests of preventive medicine, will make risk-based recommendations for a healthy lifestyle."

Challenges for the Public Health Model

While genetic testing and screening may improve health outcomes, they also bring some significant ethical, legal, and social concerns. Public health focuses largely on the health of communities, and a community approach in preventive medicine can lead to an individual's right to privacy: informed consent (i.e., access to one's genetic data for research) being outweighed by a community's goal to monitor genetic diseases effectively (Hodge 2004). What is meant by *autonomy* in such situations is indeterminate, particularly as the preventive medicine and community standards of health drive public health away from individual choice (Chadwick 2004). It may be the case that from the standpoint of promoting public health, nutrigenomics will have a greater public impact in developing countries, where public health impacts may be a decade away, than in industrialized countries (Darnton-Hill et al. 2004).

There is also a risk that once a public health initiative is under way, it will be tempting to categorize subgroups of people into risk classes for certain nutrient–gene interactions. People who belong to particular groups may find themselves labeled as being in a certain genetic category whether or not they personally carry the genetic variant. There is a high potential for discrimination, as in the cases of genetic screening programs for sickle cell anemia among African-Americans in the 1970s, BRCA testing for Ashkenazi Jews, and testing of Haitians who were identified as being at high risk for HIV/AIDS (Khoury et al. 2003). These issues about the potential for race-based stigmatization will have to be balanced against the potential benefits of targeting drugs such as BiDil to selected groups of those whom will benefit most.

To summarize, certain questions need to be answered before a genetic screening program is implemented:

- Can people accurately identify an at-risk population for targeted screening?
- Should the screening be required or optional?
- Who will have access to the screening program?
- Is there an effective and affordable treatment for the condition being screened?
- Can the screening program accomplish a public health goal?
- Does the public actually need, and is it willing to accept, the screening program?
- For genetic tests to have practical and thus public health value, they must be evaluated for such factors as sensitivity, specificity, analytical

validity, and clinical validity. These essential factors can be determined using population-based studies. However, the cost-effectiveness of genetic testing will generally depend on the value of this information to patients and society. Public health recommendations are increasingly being based on systematic evidence analysis of the benefits, harms, and costs of proposed measures. A full systematic review using a formal evidence analysis approach will probably be necessary prior to adoption by the public health systems.

- Finally, test validity is not the only issue to resolve before administering genetic tests in the public health domain. Attention should be paid to the appropriate postcounseling context and the training of physicians in genetics. It is the physician who is the patient's most important and immediate contact with the health care system.

4.4   CONCLUSION

Two principal questions apply to all health practitioner models: Is there an obligation on the part of the professional to provide these tests and/or to respond to patient requests for their professional input? and: Is there an obligation on the part of governments to provide funding for these tests?

The existence of a test does not create an obligation to use it. The answer to both questions above depends on the current state of science, numbers of trained personnel and their levels of training, cost and time considerations, and whether other feasible options are available. Taking these factors into account, it is arguable that there is currently no obligation on the part of professionals or society to provide nutrigenomic tests or to fund them. However, this could change in the future. To the extent that these tests promise significant benefits to the patient, professionals will be obliged to offer them. As a general principle, the higher the predictive value of a nutrigenomic test and the more effective known interventions are known to be, the clearer the practitioner's duty to offer the test and society's obligation to fund it.

In addition, the health professional has a duty to protect each patient from harm. Given the current public access to nutrigenomic tests, the health professional may have an obligation to help evaluate the quality of tests, discuss test results and recommended advice, and help to interpret their importance for the health of the patient. Nutrigenomic tests have already entered the marketplace, and although the market is still small, a number of factors are expected to drive its expansion in the near future. The development of nutrigenomic services raises important issues about how these tests should be administered, by whom, and with what accompanying supports.

Therefore, the following recommendations are in order. We suggest four approaches for delivering nutrigenomic services to help frame this discussion and because they are real alternatives in nutrigenomics: the consumer model, the health practitioner model, a blended model, and the public health model. Each has strengths and weaknesses that need to be evaluated. When determining which approach will provide the best combination of service delivery and accountability, a number of factors should be considered, including the current state of the science, the predictive value of the tests, the number of trained personnel available, the levels of training and funding, and the degree of oversight and accountability. Regardless of the approach chosen, there is a need for further training of health care professionals in the areas of genetics, nutritional science, and in the skills needed to link diet and health.

REFERENCES

American College of Medical Genetics. 2004. ACMG statement on direct-to-consumer genetic testing. *Genetics in Medicine* 6:60.

Arab, L. 2004. Individualized nutritional recommendations: Do we have the measurements needed to assess risk and make dietary recommendations? *Proceedings of the Nutrition Society* 63:167–172.

Biesecker, B. B., and T. M. Marteau. 1999. The future of genetic counselling: an international perspective. *Nature: Genetics* 22:133–137.

Bouwman, L. I., G. J. Hiddink, M. A. Koelen, M. Korthals, P. van't Veer, and C. van Woerkum. 2005. Personalized nurition communication through ICT application: how to overcome the gap between potential effectiveness and reality. *European Journal of Clinical Nutrition* 59:S108–S116.

Breslow, L., J. Duffy, D. E. Beauchamp, and C. L. Soskolne. 1995. Public health. In *Encyclopedia of Bioethics*, edited by W. T. Reich. New York: Macmillan (Simon & Schuster).

Bubela, T., and T. A. Caulfield. 2004. Do the print media "hype" genetic research? A comparison of newspaper stories and peer-reviewed research papers. *Canadian Medical Association Journal* 170:1399–1407.

Buchanan, A., A. Califano, J. Kahn, E. McPherson, J. Robertson, and B. Brody. 2002. Pharmacogenetics: ethical issues and policy options. *Kennedy Institute Ethics Journal* 12:1–15.

Burke, W., and J. Emery. 2002. Genetics education for primary-care providers. *Nature Reviews: Genetics* 3:561–566.

Burton, H. 2003. *Dietitians Education: Workshop Report.* London: The Wellcome Trust and the Cambridge Public Health Genetics Unit.

Castle, D. 2003. Clinical challenges posed by new biotechnology. *Postgraduate Medical Journal* 79:65–66.

Caulfield, T. A., M. M. Burgess, B. Williams-Jones, M.-A. Baily, R. Chadwick, M. Cho, R. Deber, U. Fleishing, and C. Flood. 2001. Providing genetic testing through the private sector: a view from Canada. *ISUMA* 2:72–81.

Chadwick, R. 2004. Nutrigenomics, individualism and public health. *Proceedings of the Nutrition Society* 63:161–166.

Check, E. 2003. Consumers warned that time is not yet ripe for nutrition profiling. *Nature* 426:107.

Cooper, R. S., and B. M. Psaty. 2003. Genomics and medicine: distraction, incremental progress, or the dawn of a new age? *Annals of Internal Medicine* 138:576–580.

CTV. 2006. N.B. man told to give up smoking or forget surgery. Bell Globemedia Inc. Retrieved October 2, 2006, from `www.ctv.ca/servlet/ArticleNews/story/CTVNews/20060209/smoking_surgery_060209/20060209`.

Darnton-Hill, I., B. Margretts, and R. Deckelbaum. 2004. Public health nutrition and genetics: implications for nutrition policy and promotion. *Proceedings of the Nutrition Society* 63:173–185.

DeBusk, R. M, C. P. Fogarty, J. M. Ordovas, and K. S. Kornman. 2005. Nutritional genomics in practice: Where do we begin? *Journal of the American Dietetic Association* 105:589–598.

Duncan, D. E. 2002. DNA as destiny. *Wired Magazine*: Issue 10–11, 1–4.

Eaton, C. B., P. E. McBride, K. A. Gans, and G. L. Underbakke. 2003. Teaching nutrition skills to primary care practitioners. *Journal of Nutrition* 133:563S–566S.

Flocke, S. A., and K. C. Stange. 2004. Direct observation and patient recall of health behavior advice. *Preventive Medicine* 38:343–349.

GeneWatch. 2003. The dangers of genetic testing kits. *The Observer*.

German, J. B. 2005. Genetic dietetics: nutrigenomics and the future of dietetics practice. *Journal of the American Dietetic Association* 2005:530–531.

German, J. B., C. Yeretzian, and H. J. Watzke. 2004. Personalizing food for health and preference. *Food Technology* 58:26–31.

Goldberg, J. P. 2000. Nutrition communication in the 21st century: what are the challenges and how can we meet them? *Nutrition* 16:644–646.

Gollust, S. E. 2003. Direct-to-consumer sales of genetic services on the Internet. *Genetics in Medicine* 5:332–337.

Gollust, S. E., S. C. Hull, and B. S. Wilfond. 2002. Limitations of direct-to-consumer advertising for clinical genetic testing. *Journal of the American Medical Association* 288:1762–1767.

Greendale, K., and R. E. Pyeritz. 2001. Empowering primary care health professionals in medical genetics: How soon? How fast? How far? *American Journal of Medical Genetics* 106:223–232.

Guttmacher, A. E., J. Jenkins, and W. R. Uhlmann. 2001. Genomic medicine: Who will practice it? A call to open arms. *American Journal of Medical Genetics* 106:216–222.

Hall, J. G. 2003. Individualized medicine: what the genetic revolution will bring to health care in the 21st century. *Canadian Family Physician* 49:12–13, 15–17.

Halsted, C. H. 1999. The relevance of clinical nutrition education and role models to the practice of medicine. *European Journal of Clinical Nutrition* 53(S2):S29–S34.

Hinsliff, G. 2002. Clampdown on DIY gene test kits: watchdog fears abuses. *The Observer*.

Hodge, J. G., Jr. 2004. Ethical issues concerning genetic testing and screening in public health. *American Journal of Medical Genetics* 125C:66–70.

Holtzman, N. A., and M. S. Watson. 1997. *Promoting Safe and Effective Genetic Testing in the United States: Final Report of the Task Force on Genetic Testing*. Washington, DC: National Human Genome Research Institute.

Hudak, P. L., P. McKeever, and J. G. Wright. 2003. The metaphor of patients as consumers: implications for measuring satisfaction. *Journal of Clinical Epidemiology* 56:103–108.

Human Genetics Commission. 2003. *Genes Direct: Ensuring the Effective Oversight of Genetic Tests Supplied Directly to the Public.* London: HGC.

Institute for the Future. 2002. *The Future of Nutrition: Consumers Engage with Science.* Cupertino, CA: IF.

———. 2003. *From Nutrigenomic Science to Personalized Nutrition: The Market in 2010.* Cupertino, CA: IF.

Janata, J. 2003. Piloting home tests. *The Beacon Journal.* March 18.

Jenkins, J., M. Blitzer, K. Boehm, S. Feetham, E. Gettig, A. Johnson, E. V. Lapham, A. F. Patenaude, P. P. Renolds, and A. E. Guttmacher. 2001. Recommendation of the core competencies in genetics essential for all health professionals. *Genetics in Medicine* 3:155–159.

Kapp, M. 2005. Body wise. *O, The Oprah Magazine* December: 201–202, 208–209.

Kearney, J. M., and S. McElhone. 1999. Perceived barriers in trying to eat healthier: the results of a pan-EU consumer attitudinal survey. *British Journal of Nutrition* 81:S133–S137.

Khoury, M. J., L. L. McCabe, and E. R. B. McCabe. 2003. Population screening in the age of genomic medicine. *New England Journal of Medicine* 348:50–58.

Koelen, M. A., and B. Lindström. 2005. Making health choices: the role of empowerment. *European Journal of Clinical Nutrition* 59:S10–S16.

Kolasa, K. M. 2005. Strategies to enhance effectiveness of individual based nutrition communications. *European Journal of Clinical Nutrition* 59:S24–S30.

Korf, B. R. 2002. Integration of genetics into clinical teaching in medical school education. *Genetics in Medicine* 4(6, Suppl.):33S–38S.

Kummer, C. 2005. Your genomic diet. *Technology Review* August: 54–58.

Kushner, R. F. 1995. Barriers to providing nutrition counseling by physicians: a survey of primary care practitioners. *Preventive Medicine* 24(6):546–552.

Ledley, F. 2002. A consumer charter for genomic services. *Nature: Biotechnology* 20:767.

Marteau, T. M. 1999. Communicating genetic risk information. *British Medical Bulletin* 55(2):414–428.

Marteau, T. M., V. Senior, S. E. Humphries, M. Bobrow, T. Cranston, M. A. Crook, L. Day, M. Fernandez, R. Horne, A. Iverson, et al. 2004. Psychological impact of genetic testing for familial hypercholesterolemia within a previously aware population: a randomised controlled trial. *American Journal of Medical Genetics* 128:285–293.

McCabe, L. L., and E. R. B. McCabe. 2004. Direct-to-consumer genetic testing: access and marketing. *Genetics in Medicine* 6:58–59.

Meek, J. 2002. Public "misled by gene test hype". *The Guardian.*

Moore, H., and A. J. Adamson. 2002. Nutrition interventions by primary care staff: a survey of involvement, knowledge and attitude. *Public Health Nutrition* 5:531–536.

Murray, T. H., and J. R. Botkin. 1995. Genetic testing and screening: ethical issues. In *Encyclopedia of Bioethics*, edited by W. T. Reich. New York: Macmillan (Simon & Schuster).

National Institute for Health and Clinical Excellence. 2005. *Social Value Judgements: Principles for the Development of NICE Guidance.* NIHCE. London.

Nature Medicine. 2003. Getting a grip on genetic testing. *Nature: Medicine* 9:147.

Patterson, R. E., D. L. Eaton, and J. D. Potter. 1999. The genetic revolution: change and challenge for the dietetics profession. *Journal of the American Dietetic Association* 99:1412–1420.

Peltola, K. 2002. Nutrition: communicating the message. *Nutrition Bulletin* 27:265–269.

Pray, L. A. 2005. Dieting for the genome generation. *The Scientist* 19:14.

Romain, G. 2003. Gene-tailored diets spark ethical debate. Retrieved June 11, 2003 from www.betterhumans.com/News/news.aspx?articleID=2003-11-06-1.

Rundle, R. L. 2005. Can genetic tests aid in nutrition? *Wall Street Journal Online*. March 1.

Skipper, A. 2004. The history and development of advanced practice nursing: lessons for dietetics. *Journal of the American Dietetic Association* 104:1007–1012.

Stevenson, M. 1999. Good gene hunting: commercializing safety and efficacy of home genetic test kits. *Journal of Biolaw and Business* 3:29–39.

Truswell, A. S. 1999. What nutrition knowledge and skills do primary care physicians need to have, and how should this be communicated? *European Journal of Clinical Nutrition* 53(S2):S67–S71.

U.S. Preventive Services Task Force. 2003. Behavioral counseling in primary care to promote a healthy diet: recommendations and rationale. *American Journal of Preventive Medicine* 24:93–100.

Vickery, C. E., and N. Cotugna. 2005. Incorporating human genetics into dietetics: curricula remains a challenge. *Journal of the American Dietetic Association* 105:583–588.

Vineis, P., and D. C. Christiani. 2004. Genetic testing for sale. *Epidemiology* 15:3–5.

Walker, W. A. 2003. Innovative teaching strategies for training physicians in clinical nutrition: an overview. *Journal of Nutrition* 133(2):541S–543S.

Wang, C., R. Gonzalez, and S. D. Merajver. 2004. Assessment of genetic testing and related counseling services: current research and future directions. *Social Science and Medicine* 58:1427–1442.

Yach, D., D. Stuckler, and K. D. Brownell. 2006. Epidemiologic and economic consequences of global epidemics of obesity and diabetes. *Nature: Medicine* 12:62–66.

Zimmern, R., J. Emery, and T. Richards. 2001. Putting genetics in perspective. *British Medical Journal* 322(7293):1005–1006.

5

NUTRIGENOMICS AND THE REGULATION OF HEALTH CLAIMS FOR FOODS AND DRUGS

5.1 INTRODUCTION

All emerging fields of science and technology change, however slightly or greatly, the social milieu in which basic research is done, products and services are developed, and the public is exposed to the products and services. Regulators face the difficult challenge of knowing when it is appropriate and necessary to regulate new science and technology. New science and technology, if underregulated, can needlessly expose the public to new risks. If the science and technology is overregulated, there is the potential for stifling innovation and depriving the public of the benefits of the science and technology. Regulatory authorities must always judge whether a new science and technology needs to be regulated in the first place, or whether some other option, such as voluntary compliance with government- or industry-generated guidelines, is sufficient. An example of the latter is the development of the European Nutrigenomics Organisation's development of guidelines for nutrigenomics research (Bergmann 2006; Saito et al. 2005). If the decision is to regulate, part of the challenge lies in knowing when existing regulatory frameworks can no longer be extended to new science and

Science, Society, and the Supermarket: The Opportunities and Challenges of Nutrigenomics,
By David Castle, Cheryl Cline, Abdallah S. Daar, Charoula Tsamis, and Peter A. Singer
Copyright © 2007 John Wiley & Sons, Inc.

technology. In cases where they cannot, new regulations may need to be developed if authorities decide that they are warranted.

Regardless of whether old regulatory frameworks can be extended, or new ones are required, it is a rule of thumb that regulation should be proportionate to the risks introduced by the new technology. As simple as this sounds, it is difficult to put into practice in the face of changing circumstances and the uncertainties associated with new science and technology. As we saw earlier, several uncertainties are associated with nutrigenomics. These include the current and future strength of the science, the direction in which research and service delivery will take information and biological sample management, and the potential for several simultaneous or overlapping models for service delivery. The need for regulation may be identified for any one of these issues, and it may vary significantly from country to country. However, a handful of issues in nutrigenomics have already attracted the attention of regulators.

In this chapter we begin by considering briefly three regulatory issues: genetic tests, service, and genetic antidiscrimination, to illustrate our first main point—that regulations specific to nutrigenomics do not exist. This regulatory uncertainty will in each case either persist or will be handled differently across jurisdictions. Although we have raised these as ethical issues for nutrigenomics and have discussed the potential for harm in other chapters, we note that some of the regulatory response to each concern that might apply to nutrigenomics will be driven by regulations adopted for other technologies. For example, genetic antidiscrimination regulations will be developed for the whole of human genetics, not just nutrigenomics, and direct-to-consumer nutrigenomic tests might be swept up in the regulation of an array of other direct-to-consumer tests, including paternity tests and identity confirmation tests.

Our second major point about regulation follows from the first, which is that regulations for nutrigenomics will often take the form of crossover regulations between regulatory silos that are traditionally kept more separate. This is to be expected, given that nutrigenomics is a good example of a technology involving the convergence of different subspecializations of science and technology. Convergent biotechnologies challenge regulators to adapt to new ways of thinking about regulation that arise from traditional silo-based regulation, and to use "integrated" or "horizontal" regulatory models in which different regulatory regimes also converge (Castle et al. 2006). In the case of nutrigenomics, the most pressing concerns will arise in the crossover between the regulation of foods, drugs, and the health claims made on behalf of each.

This brings us to our third general point about nutrigenomic regulation. It would be useful to have models of regulation that will guide the needs assessment for nutrigenomics regulation and the downstream development of nutrigenomics regulations should the need arise. The closest and most relevant models are in the regulation of supplements, functional foods, and nutraceuticals. These we consider in detail because they will provide us with insights that will lay the foundation for the regulation of nutrigenomics in the future. Accordingly, in the bulk of this chapter we identify and provide insight on the relationship between nutrigenomics, functional foods and nutraceuticals, and health claims.

The fourth and final general point about regulation is the following. If nutrigenomics will require crossover regulations of the sort already under consideration for supplements, functional foods, and nutraceuticals, it will be necessary to consider generically how the convergence between food and drug regulation is being addressed in different jurisdictions. Of course, this is a topic for a major research study itself, so the analysis here will be light on details in order to articulate how regulators will have to respond to the convergence trend we are identifying. We conclude the chapter with recommendations that will guide reform of existing regulation and the development of new regulations for nutrigenomics.

5.1.1 Genetic Tests, Service Delivery, and Genetic Antidiscrimination

We begin with the issue of regulating genetic tests that are administered in nutrigenomic service delivery, an issue made more complex by the Internet and direct-to-consumer provision of tests across borders (Gollust 2003). In the United States, the test kits are themselves regulated as medical devices by the Food and Drug Administration, but there is no nutrigenomic-specific regulation for the tests. In the Canadian context, there is no single formal mechanism for monitoring the marketing and delivery of genetic tests in Canada (Caulfield and Burgess 2001), let alone nutrigenomic tests. In reality, the situation is complex, since no single agency regulates the test itself, the control of the test results, the information generated from the test, or the provision of tests from non-Canadian firms. Whether and to what extent nutrigenomic tests will be regulated in Canada and the United States depends on the balance of private versus public provision of the services, whether nutrigenomics is categorized as if it were a wellness or lifestyle test or something closer to a medical diagnostic test, and on how the tests are marketed. Given that nutrigenomics is comprised of an array of complementary technologies—some for private provision, some closer to lifestyle testing than medical testing, some better suited to a medical context—it will be difficult to judge whether nutrigenomic genetic testing can be addressed within preexisting regulatory frameworks.

The regulation of genetic tests has received attention in the United Kingdom, not so much because of issues associated with the reliability and content of the test itself, but because of the service delivery model being deployed. When Sciona began offering its tests for sale directly to the public through retail outlets such as the Body Shop, it quickly attracted media attention (Meek 2002). GeneWatch and the UK Consumers' Association initiated a campaign to raise public awareness about the nutrigenomics service and how it was being marketed, hoping to raise concerns in the public's mind about the potential for these tests to harm the public. GeneWatch raised issues discussed in Chapter 3, including the potential for consumers to learn things about their health which they or their family members may not wish to know, the potential for insurers and employers to gain access to personal genetic information, and the potential for misuse of personal and genetic information in patent claims or subsequent research for which there is no consent (Meek 2002; BBC 2003) Sciona was caught in a regulatory vacuum, since at the time here were no regulations available to guide the company and to allay media, activist, and public concerns about the technology.

The effects of underregulation and negative media exposure forced Sciona to abandon its direct-to-consumer services, and other companies hoping to market in the United Kingdom, such as Genovations, also received negative media attention (*Guardian* 2003). By the spring of 2003, the Human Genetics Commission had concluded that direct-to-consumer tests ought to be discouraged because of the potential for nonconsensual testing, particularly in the case of children (Human Genetics Commission 2003), a position that was buttressed (European Group on Ethics in Science and New Technology 2003) and welcomed by GeneWatch (2003). As we noted in Chapter 3, the testing of children and adolescents is a serious ethical issue in its own right, and the possibility that harm could be exacerbated by nonconsensual testing is worrisome. The Human Genetics Commission concluded that the net effect of their recommendation "may be to reduce the availability of home sampling and mail order testing services in favour of some form of direct face-to-face contact. In our view this ideally would be with a health professional or other responsible individual who can obtain proper consent and ensure that the sample is from whom it purports to be from" (Human Genetics Commission 2003). In terms of the service delivery models we discussed in Chapter 4, the Human Genetics Commission recommendation would drive the direct-to-consumer provision of tests toward a primary health care practitioner model or perhaps a blended service delivery model.

As we discussed in Chapter 3, one of the most serious issues associated with nutrigenomics is the potential for genetic information or biological samples to be misused. The typical concern is that misappropriation and use of this information could constitute discrimination were it used to unfairly deny people access to insurance and employment. This issue is well known beyond the field of nutrigenomics, but nutrigenomics raises the question of how seriously people will take genetic privacy if the genetic test is regarded as being "just about lifestyle" or "just about food." There is, of course, no difference between a nutrigenomic test and any other, particularly with respect to providing biological samples but also with respect to those variants that are strongly associated and well characterized with disease susceptibility. Concerns about the potential for genetic discrimination have been raised since the inception of the Human Genome Project, and some jurisdictions have taken measures to protect the public. For example, the U.S. Americans with Disabilities Act was extended to protect people based on their genetic profile, and on February 17, 2005 the U.S. Senate passed The Genetic Nondiscrimination Act of 2005 (Library of Congress 2005). In the United Kingdom, a five-year moratorium that restricts the use of genetic insurers and employers was passed in 2001 between the government and the Association of British Insurers (BBC 2001). As if to highlight the variability in regulation, Canada has no regulation regarding the use of genetic information beyond what the Charter of Rights and Freedoms might prohibit.

5.2 FOOD CATEGORIES: FUNCTIONAL FOODS, NUTRACEUTICALS, MEDICINAL FOODS, AND DIETARY SUPPLEMENTS

We see a growing array of products suggested to improve our health, labeled as functional foods, nutraceuticals, medical foods, natural health products, and dietary supplements. Although these special foods are, like drugs, intended to help prevent or manage diseases, they are generally prohibited

---

Foods and Genes

Some foods are already labeled on the basis of genetic variations among people. Those who suffer from PKU, a metabolic disorder with a genetic basis, cannot safely consume products containing aspartame, so warning labels are required on products using this sweetener (Food Standards Agency 2005). Products such as Lactaid are marketed to people who suffer from lactose intolerance (the inability to digest milk sugars), the most prevalent genetic deficiency-based condition worldwide. There are margarines containing stanol esters, which can help to decrease LDL ("bad") cholesterol levels and therefore the risk of coronary heart disease in people suffering from genetic high cholesterol (Larkin 2000). There are also foods for people with such conditions as diabetes, renal problems, and celiac sprue (Sloan 2000).

from making specific claims that they treat, mitigate, or cure a disease (Kruger and Mann 2003). In some cases, foods and drugs work along similar metabolic pathways, making it hard to distinguish between nutritional and pharmacological effects. This raises issues of proper categorization, appropriate standards of scientific substantiation, and communication about health information through labeling.

As we have discussed, nutrigenomics can be expected to play a role in public health advice via general nutritional guidelines as well as to be the cornerstone of personalized nutrition. A less-well appreciated implication of nutrigenomics research is that it will act as a driver in the improvement of supplements, functional foods, and nutraceuticals. Nutrigenomics may aid in the substantiation of health claims for foods by more precisely linking the benefits of various foods with genetic differences in individuals or groups of people with similar genotypes. If people turn to nutrigenomic tests to provide advice about what to eat, it could generate more interest in the regulation of functional foods, nutraceuticals, and associated health claims, and it may in the long run necessitate the creation of a new category of health claims that couple nutrigenomics with functional foods, and nutraceuticals. This challenge will be all the greater in cases in which there is tied-selling between the nutrigenomics test and supplements, functional foods, and nutraceuticals.

Why Regulate?

Government regulators are looking for the best ways to allow consumer choice while respecting traditional diets and providing good scientific advice to the public about the risks and benefits. A major goal for regulators is to minimize the risks from products people consume to improve their health. There have been cases of herbs that were sold without restriction and later found to pose risks if consumed by certain people or in large quantities. At the same time, the food industry needs to be able to inform consumers about the health benefits of various foods and food components. Consumers need access to information about food benefits, and need protection against false or misleading claims, or those based on inadequate evidence (Mepham 2001).

There are both costs and benefits associated with regulating new areas of science and technology. On the manufacturer's side, more stringent regulations on food products claiming specific health benefits may increase their costs and the time to get a product to market (Ziker 2003). On the other hand, a clear regulatory framework for making health claims will help to stabilize the business environment for companies involved in the manufacture, distribution, and sale of food products and the provision of testing services. From the consumer viewpoint, regulations are needed to help ensure the safety and efficacy of the products while still allowing for consumer choice in the marketplace. Historically, food regulations have helped consumers by providing nutritional information while protecting the public from unsafe products and fraudulent or misleading health claims.

As we mentioned above, the challenge ahead is to judge whether regulators will be able to identify and willing to address any new risks while allowing public access to the potential benefits of nutrigenomics.

5.2.1 Functional Foods

Some components in our diets, including those not required for nutrition, may influence growth, development, and performance, and prevent or at least reduce the risk of certain diseases (Milner 2002). These bioactive or functional ingredients are believed to achieve their beneficial effects by selectively altering one or more physiological processes (Milner 2000; Hasler 2002; Kruger and Mann 2003). Functional foods can be whole or processed, enriched with additives, or modified through conventional techniques or by genetic modification. These foods are generally intended primarily for function improvement or longer-term disease risk reduction in healthy people rather than as treatments or cures for people who are unwell.

Although the concept of functional foods is widely reported to have arisen in Japan in the late 1980s, the idea that particular foods have specific health benefits has a much longer history. In the United States, Coca-Cola was marketed as a functional food at the beginning of the twentieth century, and in the late 1960s, Unilever developed spreads high in polyunsaturated fatty acids, intended to help reduce blood cholesterol levels (Westrate et al. 2002). There is a growing body of nutritional research that provides scientific backing for long-suspected associations between particular foods and specific health benefits. Although there is no universally accepted definition of functional foods, several organizations have attempted to define the evolving food category. In 1998, Health Canada proposed the definition of a functional food as: "similar in appearance to a conventional food, to be consumed as apart of the usual diet, to demonstrate physiological benefits, and/or to reduce the risk of chronic disease beyond basic nutritional functions" (Fitzpatrick 2004). The Institute of Medicine of the National Academy of Sciences defines functional foods to include foods that have been manipulated or modified to enhance their contribution to a healthful diet (American Dietetic Association 2004). Accordingly, functional foods can range from unmodified whole foods such as fruits and vegetables to modified foods that have been fortified with nutrients or enhanced with phytochemicals or botanicals, such as eggs with omega-3 fatty acids (American Dietetic Association 2004).

Some functional foods are produced either by maximizing beneficial food components or by minimizing nonbeneficial components (Arai 2002). The functional components of foods include macronutrients with specific physiological effects, essential micronutrients in intakes above and beyond daily recommendations, nonessential components with some nutritive value, and some food components with no nutritive value (Roberfroid 1999; 2000). Foods believed to have an effect beyond the relief of classical vitamin or mineral deficiencies include bran, which relieves common forms of constipation and might also have other benefits, unsaturated fatty acids, which help reduce risk for coronary heart disease; and foods high in potassium and low in sodium, which help to reduce blood pressure (Katan 1999). The (omega-3) fatty acids, probiotics (viable microorganisms that are beneficial to human health), prebiotics (nondigestible food ingredients that affect the body beneficially by selectively stimulating the growth and/or activity of beneficial bacteria in the colon), and phytochemicals are examples of functional food ingredients currently being studied for beneficial health effects (Hasler et al. 2001; Hasler 2002). The International Life Sciences Institute proposed six groups of cases illustrating the biological benefits to be derived from functional foods, which include benefits for development and growth, health, and performance: "(1) growth, development and differentiation including hormonal modulation, (2) substrate metabolism, (3) defense mechanism against reactive molecular species including free radicals and oxidants, (4) cardiovascular health, (5) gastrointestinal physiology and function, and (6) mood, behaviour and psychological functioning" (Diplock et al. 1999).

5.2.2 Nutraceuticals

As in the case of functional foods, there is no single, universally recognized definition for nutraceuticals. Health Canada describes nutraceuticals as products that have been isolated or purified from foods. They are generally sold in medicinal forms, such as pills or liquid. They have been shown to exhibit a physiological benefit or to provide protection against chronic disease. Examples of nutraceuticals include fortified dairy products, citrus fruits, and fish oils rich in omega-3 fatty acids. Such fish oils are considered to have the ability to the risk of heart disease and stroke. Nutraceuticals are often prized for their concentration and dose standardization, which makes including them in a diet easy to control and makes for more straightforward anticipation of their benefits than exists when consuming functional foods.

5.2.3 Medical or Medicinal Foods

Medical or medicinal foods, sometimes called *pharmafoods*, are designed for dietary management of a disease or condition that is closely linked to specific nutritional requirements. A medical food is not authorized to claim that it will cure, mitigate, treat, or prevent a disease. Such claims will create drug status for the product. It is only permitted to claim that a medicinal food will manage a patient's special dietary needs that exist because of a disease, as opposed to treating the disease itself (Noonan and Noonan 2004). More important, these types of foods have been formulated to be administered under the supervision of a physician.

For several decades, foods have been marketed in some countries with the express goal of helping to manage a number of specific conditions. In the United States, there are products designed to help control glucose levels and to reduce the risk of complications related to diabetes. They are sold through direct-to-home drug and health supply companies, health care agencies and institutions, and even some retail sales outlets and pharmacies (Hollingsworth 2002). One medically specific food product is the HeartBar, reported to be the first medical food to help fight heart and blood vessel disease (Skelly 1999).

5.2.4   Dietary Supplements

A dietary supplement is not sold as a conventional food but as an adjunct to one's diet. It contains such ingredients as vitamins, minerals, herbs or other botanicals, amino acids, and substances such as enzymes, organ tissues, glandulars, and metabolites. It is intended to be swallowed in the form of a tablet, capsule, powder, gel, liquid, or bar. Examples of dietary supplements are the mineral selenium and vitamin E.

5.3   HEALTH-RELATED CLAIMS ASSOCIATED WITH FOODS COMPARED TO DRUGS

Because functional foods, nutraceuticals, medical foods, and dietary supplements are believed to have physiological effects beyond those associated with typical nutrients, consumers need to be informed about the specific benefits and risks of consuming such products. Placement of a product in one of the categories above could help determine what type of health-related claims can be made in relation to a particular product. Most regulatory systems make distinctions between claims that are permitted for drugs and claims that are allowable for foods. Drug claims communicate to the consumer that the product may cure, mitigate, treat, or prevent disease. Currently, drug claims are prohibited for use on foods and dietary supplements in most jurisdictions. Food and dietary supplements can use a second category of claims, which include structure/function claims, health claims, and medical foods claims.

5.3.1 Structure–Function Claims

A *structure–function claim* is about the interaction between a food or food component and a specific function of the body, although it is not a claim to cure a particular disease. Structure–function claims describe the role of a nutrient or dietary ingredient intended to affect the structure or function in humans, and typically refer to the maintenance of healthy levels or normal functioning. They may even describe the mechanism by which a nutrient or dietary ingredient acts to maintain such structure or function. Among their claims is a description of the general well-being from consumption of a nutrient or dietary ingredient. They may go so far as to claim to alleviate a classical nutrient deficiency associated with a given disease, which is the U.S. characterization of a structure–function claim. Examples of such claims include rebalancing metabolic activities, strengthening immune function, and improving the bioavailability of nutrients (Roberfroid 1999; Ziker 2003). A model structure–function claim for dietary supplements containing echinacea would be that the herb "helps support a healthy immune system" (Hasler 2002). In the United States, structure–function claims must be accompanied by the following disclaimer: "This statement has not been evaluated by the Food and Drug Administration (FDA)." In order to distinguish it from health claims allowable for drugs, the label must also state that "this product is not intended to diagnose, treat, cure or prevent any disease" (Hasler 2002). Conventional foods can also make structure–function claims, although in these cases the FDA disclaimer is not required (Hasler 2002).

5.3.2 Health Claims

Unlike structure–function claims, health claims explicitly characterize the relationship of a particular food or food component to particular diseases or health-related conditions, and are somewhat analogous to drug claims (Ziker 2003). A model health claim associated with fiber contained in grain products, fruits, and vegetables would be: "Low fat diets rich in fiber containing grain products, fruits and vegetables reduce the risk of some types of cancer, a disease associated with many factors" (Hasler 2002).

5.3.3 Medical Food Claims

In the United States, medical foods are regulated by the Food and Drug Administration under the Orphan Drug Act (1988). It defines medicinal products as products that help to manage particular diseases that have specific nutritional requirements and allows companies to make much more specific health claims for medical food than they can for dietary supplements, while avoiding the lengthy review process required for new drugs. The products, which are required to have strong scientific backing, are intended to be taken under the supervision of a physician, but can be purchased without a prescription. There are approximately 200 medical foods on the U.S. market. They are typically sold through hospitals or pharmacies to small segments of the population (Sharpe 1998). HeartBar, a chewy food bar containing the amino acid L-arginine, was approved under the Orphan Drug Act and allowed to claim that it helped to ease the symptoms of heart disease. Other examples include products that deal with Crohn's disease, a chronic inflammation of the digestive tract.

5.3.4   Disease Risk Reduction Claims

Disease risk reduction claims say that consuming a specific food or food component(s) is associated with a reduction in the risk of disease (Roberfroid 1999; 2000). An example of a risk reduction claim would be: "Iron can help reduce the risk of anemia. This food is a high source of iron." This category of claim appears to overlap with the "prevention" component of claims permitted for drugs.

5.4   NUTRIGENOMIC INFORMATION AND THE REGULATION OF FOODS COMPARED TO DRUGS

With the completion of the Human Genome Project and the evolving field of nutrigenomics, it is now possible to identify sets of genes that modulate metabolic pathways and nutrient requirements. Traditionally, drugs have been defined as articles intended for use in the diagnosis, cure, mitigation, treatment, or prevention of disease (Milner 2000). The suggestion that certain foods or their components offer some specific health benefits may blur the distinction between food and drugs, posing a challenge for regulators. For example, there has been a debate about whether plant sterols, which are added to some margarines to help reduce LDL cholesterol, should be considered food or drugs. Classification issues are central because the decision about whether a product is a food or a drug determines its regulatory status (Milner 2002). This implies very different standards and tests for safety, efficacy, and substantiation of health claims.

These issues are even further complicated with the increasing availability of fortified foods and supplements that have lead to concerns about what levels of intake might be considered safe and how safety should be assessed.

Nutrigenomics, along with the study of functional foods and nutraceuticals, is paving the way for understanding the interactions between genetic variation and nutrition and will ultimately aid in the efficacy and safety evaluation of food components or more specifically, in the risk–benefit assessment of foods. Currently, adverse effects of food components are usually assessed by looking at both low (deficiency) as well as too high (toxicity) levels of intake (Ommen 2004). Nutrigenomics research is expected to refine the safety assessment or risk–benefit assessment of food components and their accompanying health claims. In the following section we look at the challenges facing the regulation of foods and drugs in an effort to ensure their safety. Currently, there are two main strategies for dealing with these new regulatory challenges. Products are generally categorized as either food or drugs, but the distinctions are not absolute. An active ingredient might appear in a substance called a food, but if sold in a concentrated form in a tablet, it might be considered a drug.

5.4.1   The Regulation of Foods

For food there are important safety considerations regarding the content of chemicals, toxins, and microorganisms. It seems reasonable that functional foods be regulated in the same manner as traditional foods, since they either are conventional foods or are derived from them. It is expected that people will consume functional foods for their nutrition and taste as well as for their additional health benefits. Like traditional foods, functional foods may be consumed regularly over the course of a lifetime by the general population as a regular part of the diet and in an unsupervised fashion. To make it clear to consumers that functional foods are not medicine to be taken occasionally in doses but rather, that they can be eaten daily like regular food, regulations require that products must be in food form, not in pill or capsule form (Skelly 1999). However, there are a number of problems with categorizing and regulating functional foods as traditional foods.

First, in contrast to drugs, it is widely assumed that increased consumption of foods and food ingredients is "safe" and that any physiological response will be associated with a reduction in disease risk rather than with a cure or treatment. Both of these assumptions are questionable. Functional foods are biologically active, and if too much of any substance is consumed, it can have harmful effects, particularly for people with sensitivities to such a substance or where there is a small margin between an intended intake level and a harmful dose (Kruger and Mann 2003). For example, concerns have been raised about high levels of consumption of the phytoestrogens found in soy. Although diets rich in soy have been shown to reduce the risk of cardiovascular disease and some cancers, there are worries that phytoestrogens might stimulate the growth of estrogen-dependent breast cancers.

In the case of drugs, unwanted toxicity can be minimized by a careful monitoring and adjustment of dose amounts, typically under the supervision and guidance of a health professional or through appropriate drug labeling. The consumption of substances treated as foods is unsupervised, making the problem of potential toxicity or overconsumption more difficult to regulate or monitor. As with drugs, there is a potential for adverse interactions between functional foods and other dietary components or pharmaceutical agents (Kruger and Mann 2003). One example from recent years is that of St. John's wort, a herb that was used to treat mild depression. This herb has been found to deactivate a number of drugs, including medications for heart disease, seizures, cancer, and birth control, by decreasing their levels and activity in the body (Hasler 2002). Such potential for harmful side effects is leading some to conclude that, like drugs, functional foods should be regulated to require evidence of their safety, and this safety information should be required on labels (Milner 2000; Hasler 2002). These cases indicate that consumers should discuss the use of health-promoting food products with their health care providers. The health care community also needs to be informed about interactions so that they can inform patients about possible hazards.

Finally, there is concern that the placement of functional foods and nutraceuticals in a food category might permit the passage of medicinals as food products. In the United States, for example, companies often spend years and millions of dollars getting FDA approval to market new drugs or food additives. There is a possibility that companies could position their products as medical foods, functional foods, or nutraceuticals, thus avoiding the more stringent regulatory requirements they would have to follow if their product was marketed as a pharmaceutical.

5.4.2 The Regulation of Drugs

Whereas foods are expected to be harmless, drugs are seen as powerful agents acting on our bodies, and they must pass risk–benefit analyses. Safety issues related to the use of functional foods extend beyond traditional food concerns such as contamination, to include safety concerns that could arise from their consumption. Since the consumption of functional foods and nutraceuticals carries potential risks, it seems reasonable to subject them to some kind of regulation based on a risk–benefit analysis, particularly given that they might be used over a long period and without medical supervision. If placed in a druglike category, assurance of their safety would require evidence not only of positive health benefits but also of the potential risks of adverse health effects. Protecting public health through safety assurance is critical to ensure that consumer confidence in the food supply is not jeopardized (Milner 2000). However, the stringent safety standards used for drugs may not be appropriate for what still are foods. Many functional food products may present a minimal safety risk, provided that they are manufactured under sanitary conditions; and for those functional foods that have a history of use by large populations, there is already some assurance of safety. Katan has suggested that widespread acceptance and use of certain functional foods for their health-promoting properties already gives some measure of their efficacy (Katan 1999). Having too high a safety or efficacy standard for functional foods could lead to unnecessary delays in bringing products to market for consumers, and could entail higher than necessary costs for manufacturers.

Another concern is that regulating functional foods as drugs would lead to the unnecessary medicalization of food. There are already those who feel that the growing emphasis on diet as a way of disease risk reduction is changing how people conceive of food and its relationship to the body. Further attention to the health aspects could lead to less enjoyment of food and disrupt the role that food and eating have traditionally played in social settings. Promoting food as drugs, even for the limited purpose of regulation and market promotion, could contribute to a heightened tendency to view food as medicine. Whether this is a serious worry, however, is not clear, given the likely minor modulations required for diets tailored to individuals and given that tailored diets are already a part of lifestyle management for many people. We are seeing an increasing move to diets geared to issues and health concerns that include diabetes; vegetarianism; low-fat, low-sodium, and nondairy products; and diets that avoid nuts and gluten.

5.5 FOOD AND DRUG REGULATIONS IN JAPAN, THE UNITED STATES, AND CANADA

A number of governments have begun to address the question of what benefits can be claimed for health foods and how these claims are to be communicated to the public (Mollet and Rowland 2002). They are in the process of deciding if the products should be regulated as foods or drugs. We look at regulatory responses in three countries where there has been considerable activity in the area of health-promoting foods: Japan, the United States, and Canada.

5.5.1 Japan

Japan has been a world leader in the research and development of functional foods, and in the creation of a regulatory system for health claims for foods. In 1984, an ad hoc research group began a large-scale, government-funded national project to explore the links between food and medicine (Arai 2002). Foods have three different functions: nutrition, sensory satisfaction, and physiological functioning. After nearly a century of nutritional research focused primarily on the nutritional and sensory properties of food, an aging Japanese population coupled with escalating health care costs prompted a shift of attention toward understanding and prevention of lifestyle-related diseases, including diet (Arai 2002). The ratio of people over 65 in Japan is increasing at a faster rate than in other developed countries, with one-fourth of the total population expected to fall into this age range in 20 years' time (Shimizu 2002). Early studies focused on body-defending functions and on components such as antioxidants (Arai 2002). Later research included studies based on body modulation and food allergies.

In 1991, the Japanese Ministry of Health and Welfare instituted the world's first regulatory system intended to govern the commercialization and promotion of selected functional foods for "a specific health use" (FOSHU). As part of the Nutrition Improvement Law, the legislation regulates allowable descriptions on labels regarding the effect of food on physiology. It was designed in part to prevent poorly supported and otherwise misleading advertising of food products (Shimizu 2002). There are three requirements for FOSHU approval. The first is scientific evidence of the efficacy of the food item, including evidence from well-designed clinical trials. The second is assurance of food safety. The third requirement is "the analytical determination of the effective component" (Shimizu 2002).

Since 1993, a number of foods have been approved under this system, permitting a claim to medical status not previously allowed. The claims must be limited, however, to saying that foods will help deal with a disease in its preliminary stages or will have benefits in the case of nutrient imbalances (Shimizu 2002). Fine rice was approved as the first FOSHU product, following studies at a number of medical centers. It was soon followed by hypoallergenic soybean and flour products (Arai 2002). As of 2002 there were 293 foods with FOSHU approval. The Ministry also expanded its regulatory system in 2001. Foods with nutrition function claims (FNFCs) were added to FOSHU as two subcategories of foods that can now make health claims. Twelve vitamins and two minerals have been approved as FNFCs (Shimizu 2002).

5.5.2 United States

The U.S. Food, Drug and Cosmetics Act treats foods and drugs differently. For food, product safety has long been the key issue. The original act banned food containing "any added poisonous or deleterious substance which may render it injurious to health" (Kruger and Mann 2003). However, premarket approval of food was not required, and the government bore the burden of proof if a food was suspected of being adulterated or otherwise unsafe for human consumption. In 1958, a Food Additives Amendment required that product safety be demonstrated prior to marketing additives, thereby shifting the burden of proof for safety from the government to the manufacturer.

Although there must be a reasonable certainty of no harm from foods and dietary supplements, approval for drugs involves a risk–benefit analysis that allows for certain trade-offs. FDA approval is required for the sale of drugs, based on proof of safety and efficacy. Premarket approval and postmarket surveillance are required, and the burden of proof for safety and efficacy is on the manufacturer. Since 1990, a number of important legislative acts were introduced in the United States in response to the growing body of scientific evidence linking the consumption of certain foods to particular health outcomes. They led to a loosening of requirements for scientific substantiation of health-related claims for many food items (Hasler 2002). The first of these, the 1990 Nutrition Label Education Act, required the FDA to establish regulations requiring a number of foods to have specific information on labels. It also established rules for making claims about content and disease prevention for nutrients in foods (Milner 2000; Shimizu 2002).

There is currently no statutory definition of functional foods in the United States, and the regulatory status of food products is determined by their intended use (Milner 2002). In 2000, the United States issued a federal regulation intended to clarify structure–function claims involving the relationship between a food component and a disease or health-related condition (Verschuren 2002). Nutrient content and structure–function claims are clearly defined in regulation, but FDA approval is not required. Health claims, however, must meet a standard of significant scientific agreement and be supported either by a review of existing scientific literature or by an authoritative statement of a scientific body of the U.S. government or the National Academy of Sciences. Fifteen claims linking a food substance to a health-related condition have been approved, covering such products as oat soluble fiber, soluble fiber from psyllium seed husk, soy protein, and margarines fortified with plant sterols and plant stanols.

The 1994 Dietary Supplement Health and Education Act (DSHEA) created a new framework for the regulation of such dietary supplements as vitamins, minerals, herbs and other botanicals, and amino acids. These food ingredients were identified in the act as being "for use by man to supplement the diet by increasing the total dietary intake" (Kruger and Mann 2003). This legislation also created mechanisms to deal with safety issues, the regulation of health claims, and the labeling of dietary supplements (Milner 2000). Although dietary supplements were recognized to be foods, they were placed in a new category of food "which must not be represented for use as a conventional food or as a sole item of a meal or the diet" (Kruger and Mann 2003). To distinguish dietary supplements clearly, they must be presented in tablet, capsule, powder, or liquid form (Kruger and Mann 2003). One of the features of this new class of foods is that unlike new food additives, there is no premarket approval requirement (Kruger and Mann 2003). Making health claims, nutrient content claims, or structure/function claims for dietary supplements is allowed.

The FDA Modernization Act, introduced in 1997, amended the Federal Food, Drug and Cosmetic Act to allow health claims not preauthorized by the FDA if the claims are based on authoritative statements of government agencies responsible for protecting the public health, such as the Centers for Disease Control and the National Academy of Sciences or the National Institutes of Health (Milner 2000; Hasler 2002). Claims linking the consumption of particular foods with health outcomes must be supported by the totality of scientific evidence and be the subject of significant scientific agreement. To satisfy this condition, a body of supporting evidence must exist, which includes evidence from well-designed epidemiological, laboratory, and clinical studies and expert opinions from a reputable body of independent scientists (Hasler 2002).

5.5.3 Canada

Canada's Food and Drug Act regulates the sale and advertising of foods and drugs. The food regulations under that act were originally developed to ensure that food sold was safe and nutritious and to protect the public from fraudulent claims (Milner 2002). Food is defined as "any article manufactured, sold or represented for use as food or drink for human beings, chewing gum and any ingredient that may be mixed with food for any purpose whatsoever" (Canada Food and Drug Act 1985). A drug is defined as "any substance or mixture of substances manufactured, sold or represented for use in (a) the diagnosis, mitigation or prevention of a disease, disorder or abnormal physical state, or its symptoms, in humans or animals; (b) restoring, correcting or modifying organic functions in human beings . . ." (Canada Food and Drug Act 1985).

In Canada, more than half of consumers regularly take vitamins and minerals, herbal products, homeopathic medicines, and similar substances, a category the federal government describes as natural health products. In January 2004, Canada began to phase in its Natural Health Products Regulations, a process that will take six years. The stated goal of the regulations is "to provide Canadians with ready access to natural health products that are safe, effective, and of high quality, while respecting freedom of choice and philosophical and cultural diversity." The regulations cover a wide range of substances, including vitamins, minerals, amino acids, essential fatty acids, homeopathic preparations, and a number of substances used in traditional medicines.

In these regulations these products will be considered as drugs, since it is permitted to make stronger claims for these products than those made for foods. The claims can be for the treatment or prevention of disease, restoring or correcting organic functions, or modifying those functions to maintain or improve health. The standards of evidence for safety and for a health claim will correspond to the strength of the claim. For example, a disease-related claim will now require the submission of justifiable scientific data. There is also an adverse reaction reporting system. In addition to regulating safety, the Canadian government is also regulating efficacy. The regulations cover the manufacture, packaging, labeling, storage, importation, distribution, and sale of products. They also cover what health claims can be made. However, the regulations do not cover conventional foods, so would not deal with some of the same substances contained in products sold as foods.

The Health Products and Food Branch of Health Canada has attempted to enable the use of health and risk reduction claims for foods by exempting from the provisions of the act related to drugs, foods bearing diet-related claims, provided that certain conditions are met. Based on an extensive review of the scientific basis of proposed claims, Health Canada has proposed five generic diet-related claims for adoption in Canada: sodium and hypertension, calcium and osteoporosis, saturated and trans fat and cholesterol and coronary artery disease, fruits and vegetables and cancer, and sugar alcohols and dental caries. One such allowable claim is: "A healthy diet containing foods high in potassium and low in sodium may reduce the risk of high blood pressure, a risk factor for stroke and heart disease" (Gorman 2002).

Diets rich in fruits and vegetables may reduce the risk of some types of cancer and other chronic diseases. National Cancer Institute.
         —Label on bananas sold in a Toronto supermarket in 2004

Proposed safety standards for evaluating foods with health claims include product safety, which involves a reasonable assurance that there are no adverse health effects associated with consumption of the food; validity of the claim as determined by demonstration of the efficacy of the product, by establishing a causal link between the health effect claimed and the consumption of the food or food component; and quality assurance, in the form of evidence that foods bearing health claims can identify, measure, and maintain a consistent level of the bioactive substance that ensures efficacy and do so without jeopardizing safety (Fitzpatrick 2004). For product-specific claims involving prevention or structure/function claims, the product would be subject to the act's provisions regarding drugs.

5.6 CONCLUSION

Nutrigenomics raises a number of issues for regulators, but nutrigenomics-specific regulations do not exist. Having discussed the regulation of genetic tests, service, and genetic antidiscrimination, we turned our attention to the area where we anticipate the greatest amount of crossover between nutrigenomics and emerging regulations for supplements, functional foods, and nutraceuticals. These areas in the food industry are already difficult to regulation, and the problem may grow more complex since they can be expected to be conjoined with nutrigenomics. The reason is simple and obvious—as the bioactivity of foods and food components is better characterized, the potency of nutritional interventions possible in nutrigenomics rises. Hence, the way in which supplements, functional foods, and nutraceuticals are regulated will provide early models of regulation from which insights can be drawn for the future regulation of nutrigenomics. Finally, we turned to the question about whether there will be further lessons to be drawn from the intersection of food and drug regulation, particularly since the potential for food to have pharmaceutical-like bioactivity will need proper regulation. Questions will be raised about toxicity issues in nutrigenomic dietary interventions, and in a different vein it might be asked whether the problem for regulators is that all food is bioactive in some sense or another, making the concept somewhat misleading.

Accordingly, we can recommend that in addition to ongoing scrutiny of the provision of genetic tests to the public, delivery of nutrigenomic services, and mindfulness of the potential for genetic discrimination, regulators should be cognizant of changes in the food industry. Some changes are exogenous to developments in nutrigenomics, but in other cases interest in the bioactivity of food will be driven by interests in nutrigenomics or will respond independently to nutrigenomics activity. The development of more potent nutritional interventions, and the possibility of tied-selling, means that regulators will have to consider the regulation of nutrigenomics at the same time as they consider the regulation of foods and new food products. Regulations may be needed to help ensure the safety and efficacy of products recommended as a result of nutrigenomic testing, and to ensure the validity of health claims while allowing for consumer choice in the marketplace.

This may mean a difficult road for regulators, since it might entail considerable innovation in how new science and technology is regulated (Castle et al. 2006). New regulatory concepts, definitions, and standards might have to be developed to handle the regulation of nutrigenomics, and new regulatory processes might be needed to coordinate, for example, interagency collaboration and communication. Perhaps the greatest challenge is to overcome what is variously called *silo-based* or *vertical regulation* for *integrative* or *horizontal regulation* (Lyall and Tait 2005; Castle et al. 2006). This shift in regulation will require shifts in regulatory structures and processes that may have far-reaching effects in some jurisdictions, some of which may change what is meant by regulation. Some might doubt that recommendations for such sweeping reforms in regulation are warranted, but we have argued throughout this and other chapters that nutrigenomics is sufficiently complex a field that its regulation will be correspondingly complex.

REFERENCES

American Dietetic Association. 2004. Position of the American Dietetic Association: Functional foods. *Journal of the American Dietetic Association* 104:814–826.

Arai, S. 2002. Global view on functional foods: Asian perspectives. *British Journal of Nutrition* 88(Suppl. 2):S139–S143.

Bergmann, M. M., M. Bodzioch, M. L. Bonet, C. Defoort, G. Lietz, and J. M. Mathers. 2006. Bioethics in human nutrigenomics research: European Nutrigenomics Organisation (NuGO) workshop report. *British Journal of Nutrition* 95:1024–1027.

BBC. 2003. New controls on genetic tests. Retrieved June 20, 2006, from www.bbc.co.uk/1/hi/health/2723579.stm.

———. 2001. "Moratorium" on genetic data use. Retrieved February 20, 2006, from http://news.bbc.co.uk/2/hi/business/1615397.stm.

Canada Food and Drugs Act. 1985. Food and Drugs Act (R.S. 1985, c. F-27). Canadian Department of Health. Retrieved 2003, from http://laws.justice.gc.ca/en/F-27/text.html.

Castle, D., R. Loeppky, and M. Saner. 2006. Convergence in biotechnology innovation: case studies and implications for regulation. Retrieved June 20, 2006, from www.gels.ca.

Caulfield, T. A., and M. M. Burgess. 2001. Providing genetic testing through the private sector: a view from Canada. *ISUMA* 2:72–81.

Diplock, A. T., P. J. Aggett, M. Ashwell, F. Bornet, E. B. Fern, and M. B. Roberfroid. 1999. Scientific concepts of functional foods in Europe: Consensus document. *British Journal of Nutrition* 81:S1–S27.

European Group on Ethics in Science and New Technology. 2003. Statement by the European Group on Ethics in Science and New Technologies on advertising genetic tests via the Internet. European Commission. Retrieved February 20, 2006, from `http://europa.eu.int/rapid/pressReleasesAction.do?reference=IP/03/273&format=HTML&aged=1&language=EN&guiLanguage=en`.

Fitzpatrick, K. C. 2004. Regulatory issues related to functional foods and natural health products in Canada: possible implications for manufacturers of conjugated linoleic acid. *American Journal of Clinical Nutrition* 79:S1217–S1220.

Food Standards Agency. 2004. Aspartame 2005. Retrieved June 20, 2006, from `www.food.gov.uk/safereating/chemsate/additivesbranch/sweeteners/55174`.

GeneWatch. 2003. GeneWatch UK welcomes advice to ministers to regulate genetic tests. Retrieved February 20, 2006, from `www.genewatch.org/press%20Releases/pr38.htm`.

Gollust, S. E. 2003. Direct-to-consumer sales of genetic services on the Internet. *Genetics in Medicine* 5:332–337.

Gorman, D. 2002. Health Canada's regulatory initiative regarding foods with health claims. *Canadian Journal of Public Health* 93:325–327.

Green, M. R., and F. van der Ouderaa. 2003. Nutrigenetics: Where next for the foods industry? *Pharmacogenomics Journal* 3:191–193.

Guardian. 2003. Gene test to help you beat death sparks row on ethics. Retrieved September 21, 2003, from `http://observer.guardian.co.uk/uk_news/story/0,6903,877779,00.html`.

Hasler, C. M. 2002. Functional foods: benefits, concerns and challenges—a position paper from the American Council on Science and Health. *Journal of Nutrition* 132:3772–3781.

Hasler, C., A. Moag-Stahlberg, D. Webb, and M. Hudnall. 2001. How to evaluate the safety, efficacy, and quality of functional foods and their ingredients. *Journal of the American Dietetic Association* 101:733–736.

Hollingsworth, P. 2002. Developing and marketing foods for diabetics. *Food Technology* 56:38–44.

Human Genetics Commission. 2003. *Genes Direct: Ensuring the Effective Oversight of Genetic Tests Supplied Directly to the Public.* London: HGC.

Katan, M. B. 1999. Functional foods. *Lancet* 354:794.

Kruger, C. L., and S. W. Mann. 2003. Safety evaluation of functional ingredients. *Food Chemistry and Toxicology* 41:793–805.

Larkin, M. 2000. Functional foods nibble away at serum cholesterol concentrations. *Lancet* 355:555.

Library of Congress. 2005. Genetic Information Nondiscrimination Act of 2005. Retrieved February 26, 2006, from `http://thomas.loc.gov/cgi-bin/query/z?c109:S.306.IS`.

Lyall, C., and J. Tait. 2005. *New Modes of Governance: Developing an Integrated Policy Approach to Science, Technology, Risk and the Environment.* Aldershot, Hampshire, England: Ashgate.

Meek, J. 2002. Public misled by gene test hype: scientists cast doubt on "irresponsible" claims for checks offered by Body Shop. *The Guardian*, March 12.

Mepham, B. 2001. Novel foods. In *The Concise Encyclopedia of the Ethics of New Technologies*, edited by R. Chadwick. San Diego, CA: Academic Press.

Milner, J. A. 2000. Functional foods: the US perspective. *American Journal of Clinical Nutrition* 71:1654S–1659S.

———. 2002. Functional foods and health: a US perspective. *British Journal of Nutrition* 88:S151–S158.

Mollet, B., and I. Rowland. 2002. Functional foods: at the frontier between food and pharma. *Current Opinion in Biotechnology* 13:483–485.

Noonan, W. P., and C. Noonan. 2004. Legal requirements for "functional food" claims. *Toxicology Letters* 150:19–24.

Roberfroid, M. B. 1999. Concepts in functional foods: a European perspective. *Nutrition Today* 43:162.

———. 2000. Concepts and strategy of functional food science: the European perspective. *American Journal of Clinical Nutrition* 71:1660S–1664S, 1674S–1675S.

Saito, K., S. Arai, and H. Kato. 2005. A nutrigenomics database: integrated repository for publications and associated microarray data in nutrigenomics research. *British Journal of Nutrition* 94:493–495.

Sharpe, R. 1998. Obscure regulation helps sales pitch for new food bar. Natural Healthline News. Retrieved February 20, 2003, from `www.naturalhealthline.com/newsletter/HL981215/foodbar`.

Shimizu, T. 2002. Newly established regulation in Japan: foods with health claims. *Asia Pacific Journal of Clinical Nutrition* 11:S94–S96.

Skelly, L. 1999. Functional foods: over-the counter medicine in a meal. *Healthinform: Essential Information on Alternative Health Care* 5:1.

Sloan, A. E. 2000. The top ten functional food trends. *Food Technology* 54:33–62.

Verschuren, P. M. 2002. Functional foods: scientific and global perspectives. *British Journal of Nutrition* 88:S125–S130.

Westrate, J. A., G. van Poppel, and P. M. Verschuren. 2002. Functional foods, trends and future. *British Journal of Nutrition* 88:S233–S235.

Ziker, D. 2003. Regulating functional foods: pre- and post-market strategy. Duke Law and Technology Review. Retrieved June 20, 2004, from `www.law.duke.edu/journals/dltr/articles/2002dltr0024.html`.

6

NUTRIGENOMICS: JUSTICE, EQUITY, AND ACCESS

6.1 INTRODUCTION

At the same time that we saw enormous advances in medical science and technology, particularly during the past 50 years, health inequities within and between countries continue to persist and even to widen. These inequities result from a number of causes, including widening economic disparities, rapid population growth, the emergence of new infectious diseases, escalating ecological degradation, and warfare (Benatar et al. 2003). There is concern that existing disparities in both mortality and morbidity in industrialized countries and between industrialized and developing countries could be increased either by a selective implementation or an inappropriate introduction of biotechnologies. This would further widen what is now referred to as the *health genomics divide* (Singer and Daar 2001). These concerns include the science of nutrigenomics. In light of the growing research interest in diet–gene–health interactions and the increasing development and promotion of related technologies and products, it is worth exploring what effect nutrigenomics may have on existing health inequities. By *health inequities* we mean differences in health between populations which are believed to be unfair, unnecessary, avoidable, and therefore reducible (Whitehead 1992; Braveman et al. 2001). A commitment to promoting health equity would involve efforts to reduce these gaps in health and the provision of health care.

The science of nutrigenomics is leading to the creation of tools and information that make possible increasingly tailored dietary advice on the basis of food–gene interactions. This dietary advice can be tailored to individuals or to subgroups of the general population. When tailored to individuals, it would involve personalized genetic testing for individuals who have sought the service themselves or who have had it recommended to them by a primary health care practitioner. When tailored to groups, in the case of nutrigenomics delivered through a public health initiative, individualized genetic screening might be substituted for by group-based nutritional advice, whether offered en masse for an entire country or targeted to particular ethnic groups. We examine both of these approaches for their potential impact on health inequities in both industrialized and developing countries.

There are many costs associated with nutrigenomics, including but not limited to the expense of conducting primary research in a country, developing applications with the support of private industry, administering genetic individual genetic tests, and shouldering secondary costs associated with genetic counseling services. Given the cost of taking nutrigenomics from a research concept to a technology that works for people, we anticipate a growing divide between those who can afford access and those who cannot. Based on the current state of the science, it is too early to make firm predictions about whether nutrigenomics will increase or decrease domestic or global health inequities. Assuming that public and private investment

The not-for-profit Genetic Alliance describes itself as an international support, education, and advocacy organization for all those living with genetic conditions. Founded in 1986, the Alliance (www.geneticalliance.org) is made up of over 600 advocacy, research, and health care organizations, which support millions of people with genetic conditions. The Alliance seeks to promote research into accessible technologies, to foster the integration of genetic advances into quality and affordable health care, to educate the public and health professionals, to encourage supportive public policies, to identify solutions to new problems, and to foster understanding of genetics and related policy issues.

in the science are not for naught and that the science will have significant beneficial applications, its distribution among all people is a going concern. Take, for example, the situation in industrialized countries, where nutrigenomics tests and accompanying interventions could lead to some increases in health disparities, at least in the short term, given the limited number of people who can receive such services. These people would be the early adopters we commented on in Chapter 4, who tend to be educated and affluent. On the other hand, targeted population-based public health initiatives based on nutrigenomics research have the potential to produce modest reductions in health inequities in the longer term both domestically and globally. This would be the case if the science focuses on conditions involving genetic variants that are disproportionately high in some disadvantaged subpopulations. Public health initiatives for the provision of genetic tests in medical contexts can recoup the upfront costs, but any program that requires amortization of public resources requires strong justification and political will (Morgan et al. 2003). Nutrigenomics would be no different.

6.2   INDUSTRIALIZED COUNTRY CONTEXT

6.2.1   Individualized Nutrigenomic Testing

According to consumer trend forecasters such as the Institute for the Future, the emerging nutrigenomics market, which is currently a phenomenon restricted largely to the industrialized countries, is likely to be driven by a relatively small group of consumers. Members of this group are typically female, university educated, between the ages of 50 and 64, and are willing and able to pay several hundred dollars for such a test. Their purchasing preferences include organic food products, they are proactive health information seekers, and they are more likely than the average person to experiment with alternative approaches to medical treatment. On the commercial side, the marketing and promotion of nutrigenomic tests and products as wellness tools appears to be part of a push to build a market in designer food products including supplements, nutraceuticals, and functional foods. Some companies that sell nutrigenomic test services are working with nutraceutical companies to develop health programs that involve the use of micronutrients and botanicals. Interleukin Genetics, for example, is a nutrigenomics test provider that is also developing nutritional supplements for a subset of inflammatory conditions and chronic diseases. They have partnered with Alticor for the provision of genetic tests under the Gensona brand, and have partnered with Nutrilite to develop products for consumers (Interleukin Genetics 2006).

Biotechnology company Interleukin Genetics Inc. (www.ilgenetics.com) describes itself as a soon-to-be "personalized health" company and specializes in understanding the genetics of inflammation. Inflammation is a bodily defense mechanism, and is involved in response to injury, infection, autoimmunity, and other insults. This may affect a person's development of disease, and in many cases, has roots in genetic variation. (Interleukin found that variations in the Interleukin-1 gene increases the risk of a heart attack.) The company investigates how nondrug interventions such as nutritional supplements can ameliorate genetically influenced risks. In 2000, Interleukin Genetics discovered an association between a genetic variation and a disease, and patented it as a "drug target" in 2001. This allows the company to research the gene for the development of medicines. Interleukin Genetics has an association with a commercial nutritional supplements company and is developing a cardiovascular risk assessment test for that company.

Assuming that these consumer and commercial trends continue, we may be heading toward a nutrigenomic marketplace that provides boutique services and products for, and purchased by, a niche clientele with appropriate discretionary income. Because nutrigenomics services are not currently publicly funded in any of the jurisdictions in which they are delivered, no issues are raised about competition for scarce health care resources. Similarly, it may also be suggested that the science is not sufficiently well established to create any obligations to provide these services universally. To the extent that the focus of nutrigenomics testing shifts to disease prevention or amelioration rather than wellness promotion, and as increases in genetic susceptibility and susceptibility-reducing interventions are better understood, differences in access to tests will raise equity issues. In other words, when nutrigenomic testing services are known to actually improve health outcomes, their contribution to health inequities may become more visible if there is no equity of access. Increased demand for such services may in turn generate interest in coverage for nutrigenomics being offered by health insurance.

6.2.2 Population-Based Nutrigenomics

Significant health disparities exist between different social groups in industrialized countries. Minority populations in the United States have disproportionately high rates of many chronic diseases, including type 2 diabetes, obesity, cardiovascular disease, and some cancers, compared to diseases rates in the broader population. For example, African-American men have a 60 percent greater risk of developing prostate cancer than do Caucasian men, and 50 percent of adult Pima Indians in the United States have type 2 diabetes, compared to 6.5 percent of adult Caucasian Americans (University of California–Davis 2003). Although these disparities are due to a large number of factors, including socioeconomic status, diet, culture, behavior, and lack of access to health care, genetic differences are also believed to play a role.

Despite the significant ethical and legal issues associated with thinking about disease as having a biological basis in race or ethnicity, research is nevertheless under way (Fine et al. 2005). This issue has already been encountered in the case of pharmacogenomics (Soo-Jin 2005). Western governments are starting to recognize that nutrigenomics research geared toward the health needs of minority populations has the potential to help reduce these health inequities. Contemporary research programs have to contend with the legacy of some research programs in the past that involved the mistreatment of research subjects, the effects of which are still being debated with respect to the ability to recruit minority research subjects (Freimuth et al. 2001; Wendler et al. 2006). The National Center of Excellence in Nutritional Genomics in California received a U.S. $6.5 million grant from the National Institutes of Health to explore the links between diet, genes, and disease in minority populations, with the goal of helping to reduce health disparities between these groups and the general population (University of California–Davis 2003). The center is attempting to identify genetic variations that increase susceptibility to disease in minority populations and to study how diets can be tailored to reduce this susceptibility. These types of studies could, of course, be very relevant to industrialized and developing country populations that have populations that are comparably diverse. It must be noted, however, that knowledge about health disparities and correction action are two different things. There remains in the United States, for example, significant differences in access and provision of medical services.

6.3 DEVELOPING COUNTRY CONTEXT

6.3.1 Individualized Nutrigenomic Testing

Although there are serious health inequities between subpopulations in industrialized nations, the greatest inequities in health status and disease burden are between industrialized and developing nations. Two billion people still do not have access to low-cost essential medicines (Fakuda-Paar et al. 2001). Much of the research in health genomics has been guided by the priorities of the industrialized countries in which these technologies have developed: the 10/90 gap. However, a number of groups and organizations have recently turned their attention to exploring how these technologies might be deployed to improve health in developing countries. This subject is explored in the World Health Organization's (WHO) report on genomics and world health, and the *Top 10 Biotechnologies for Improving Health in Developing Countries* report by the University of Toronto's Joint Centre for Bioethics (Gwatkin 2000; World Health Organization 2002; Daar et al. 2002).

To evaluate how nutrigenomics testing might be used to reduce global health inequities, we have adapted and modified the selection criteria used in the *Top 10 Biotechnologies* report (Daar et al. 2002). Our five criteria are health impact, disease burden reduction, appropriateness of the technology, time frame, and overall social harm–benefit ratio.

Health Impact: To What Extent Might Nutrigenomics Testing Lead to Improvements in Health?

It is widely accepted that science and technology plays an essential role in improving health in developing countries (Annan 2003). This is demonstrated by the effective implementation of preventive strategies such as genetic population screening programs in reducing the impact of congenital or genetic disorders. At least 7.6 million children are born annually with a severe congenital or genetic disorder, and a total of about 70 percent of these disorders can be treated or avoided when prevention services are in place (Alwan and Modell 2002; March of Dimes 2006). Developing countries have also utilized genomics technologies to introduce affordable and applicable tests for tuberculosis, hepatitis C, HIV, and malaria (Acharya et al. 2004). However, developing countries now face a fairly recent health challenge, the global burden of chronic diseases such as cardiovascular disease and diabetes. Almost 85 percent of this disease burden presently occurs in developing or low-resource countries and is expected to grow in the next two decades (Alwan and Modell 2002; Darnton-Hill et al. 2004; World Health Organization 2006). Although infectious diseases are currently the leading cause of high mortality rates in developing countries, chronic diseases are contributing significantly to what is called the *double burden of disease*. Crippled by this double burden, developing countries are in desperate need of disease prevention strategies coupled with nutritional intervention approaches in an effort to deal with the health challenges posed by genetic and diet-associated diseases. Unlike industrialized countries, the rising burden of chronic diseases in developing countries has received inadequate attention (Beaglehole and Yach 2003).

If harnessed adequately, the science of nutrigenomics has the potential to develop new technologies, treatments, and preventive programs in an effort to reduce the prevalence of chronic diseases in developing countries. Since the underlying science is still evolving, however, it is too early to say what impact nutrigenomics testing will have in reducing health inequities in developing nations. There is concern that the focus on genomics-based technologies may, if not applied appropriately, divert scarce and desperately needed resources away from more conventional public health strategies, such as the provision of clean water, safe and adequate food supplies, and proper sewage disposal (Pang 2003). However, the distinction between genomics/biotechnology and "traditional" measures is artificial in many cases: Vaccines, for example, span both domains. Health care providers will need to evaluate how nutritional advice from nutrigenomics testing can improve on conventional nutritional information in lower-resource countries.

Disease Burden Reduction: Does Nutrigenomics Testing Address the Most Pressing Health Needs of Those Living in Developing Countries?

The most pressing health needs in developing countries currently lie in the treatment and prevention of infectious diseases. However, chronic noncommunicable diseases are becoming increasingly important, and their incidence and prevalence are expected to rise. India already has the highest number of diabetics in the world (Yach et al. 2004). Susceptibility to highly prevalent diseases such as diabetes, hypertension, coronary heart disease, and cancer are to a large extent, genetically determined. Unhealthy diets are just one of the many leading risk factors of these diseases. The implementation of genetic tests to determine the impact of common genetic variation on nutritional requirements and risk of adverse health events may prove to be beneficial in the management of chronic disease in developing countries.

So far, nutrigenomics testing services do not have an appropriate place in developing countries because they do not help to meet the most urgent health needs at this time. This point is emphasized by experts in the field of nutrigenomics who proclaim that "we are not ready to carry out genetic screening programs for common diseases at the population level or to estimate individual risk based on genetic variation" (Simopoulos 2004). Instead, the real challenge at the moment for most developing countries is to meet basic nutritional needs across the populace rather than to tailor diets to individuals. In many cases, the recommended interventions in the form of health-promoting natural foods and supplements are often unavailable to the populace. In all of this, it is important to bear in mind that medical advances, although playing some role in reducing health inequities, are overshadowed by the importance of economic and social policies designed to improve basic living conditions through the provision of adequate housing, better education, clean drinking water, and sanitation, among other things (Benatar et al. 2003).

Appropriateness of the Technology: Would Nutrigenomic Services be Affordable for Lower-Income Populations, and How Well Would They Fit into Local Health Care Settings?

Any consideration of the suitability of the technology for meeting local health needs must be sensitive to social, cultural, political, economic, and environment factors as well as past experiences, both positive and negative, of the introduction of, or attempt to introduce, new technologies in developing countries. Most developing countries lack the financial resources, research tools, scientific knowledge, trained personnel, and health and regulatory infrastructures required for the safe, effective, and affordable delivery of nutrigenomics services. About half the world's more than 6 billion people live on less than U.S. $2 per day (World Bank 2001). With testing services currently costing in the range of U.S. $500, nutrigenomics services are currently inaccessible to most people because they are available only commercially and not through the health care system. If nutrigenomics tests were to be implemented in developing countries through population screening programs, there may not be a significant financial disparity in access to genetic testing, for genetic tests need to be done only once, and basic analytical platforms of genotyping are now becoming relatively inexpensive (Alwan and Modell 2002).

Time Frame: Is It Likely That Nutrigenomic Tests Can Be Developed and Implemented in a Relatively Short Period in Developing Countries?

The most immediately applicable, affordable, and effective application of genetic knowledge is for diagnosis and prevention, and as noted earlier, prevention is a priority for developing countries in the management of health and in reducing the burden of disease (Alwan and Modell 2002). However, similar to industrialized countries, developing countries presently lack the required resources to implement nutrigenomics services safely and effectively. The majority of developing countries currently do not have the necessary technological and human resource capacity to provide nutrigenomics services, including the sophisticated screening tools, powerful software, and Internet facilities required for data analysis and databases. In some countries there is inadequate scientific knowledge and a shortage of skilled personnel such as trained technicians and genetic counsellors.

Were scientific developments to occur rapidly in the industrialized countries in which the research is taking place, it would still take time and a great deal of concerted effort to share the fruits of this research with needier countries; and it is not clear that the will to do this would exist in any case. The poor obviously have very little purchasing power. Where there is little market incentive to sell tests and products such as supplements and nutraceuticals, and where the emphasis is on natural foods rather than drugs, the nutrigenomics industries will have little commercial interest in promoting their services and products.

Overall Social Harm/Benefit Ratio: Do the Overall Benefits of Nutrigenomic Testing Outweigh the Overall Harm for Developing Country Populations?

Although, on balance, nutrigenomics tests are not an appropriate technology for developing countries at this time, it is important to recognize the ways in which these services could help to meet developing country health needs. First, prevention is a high priority in developing countries, both as a way of reducing morbidity and mortality, and as a means of saving scarce health care dollars. Nutrigenomics testing, with its emphasis on prevention and early intervention, could help in this respect. Second, food-based interventions can be a much more affordable alternative to drugs in cases where substitution is possible. "Expenditures on medicines can represent up to 66% of total health spending in developing countries and could be a major cause of household impoverishment, as 50–90% of such expenditures are out-of-pocket expenses" (Luis 2002). Third, the emphasis on food-based interventions as a way of promoting health and preventing disease makes nutrigenomics compatible with traditional health practices in some developing countries. In some African countries up to 80 percent of the population depends on traditional medicine, and in India the number is around 65 percent (Bhardwaj et al. 2003). These and other non-Western traditions of medicine typically place greater emphasis on food-based products as a way of meeting health care needs.

Population-Based Nutrigenomics

Individualized nutrition tailored to individuals through the use of genetic tests currently has little health value for developing countries. However, population-based nutrigenomics research may be able to reduce health inequities in the longer term. This could be done if developing countries can adapt the technology to their needs and if industrialized countries assist developing countries with information and technology transfers.

A growing epidemic of noncommunicable diseases on top of long-standing communicable diseases is giving rise to the double burden of disease in many developing countries. For example, China, Malaysia, and Thailand suffer from a double burden of communicable and lifestyle diseases (World Health Organization 2002). In India, the burden of chronic diseases just exceeds that of communicable diseases. Obesity has become a serious problem in Asia, Latin America, and parts of Africa despite a widespread problem of undernutrition (World Health Organization 2002; Eaton 2003). Nutrition-related chronic conditions such as cardiovascular disease, obesity, and diabetes are now increasing even in poorer nations (Zimmet et al. 2001; Chopra et al. 2002; Powles and Day 2002; Reddy 2002; Eaton 2003). According to recent projections, by the year 2020 noncommunicable diseases will account for over 60 percent of the disease burden and mortality in developing countries (Caballero 2001; Eaton 2003). Changes in local dietary patterns, the influence of advertising, the global spread of Western diets (including highly processed foods high in fat and low in unrefined carbohydrates), and a widespread reduction in physical activity all contribute to this global shift in diet and lifestyle-related disease patterns (Eaton 2003).

Genetic variability is also known to play a role in increased susceptibility to these diet-related diseases. Studies on genetic responses to variations in diet indicate that some genotypes raise cholesterol levels more than others (Eaton 2003). For example, extremely high frequencies of the ApoE4 allele, which in part can account for the high prevalence in cardiovascular disease, have been found in African and Asian subpopulations such as New Guineans (35 percent) and Nigerians (30 percent) compared to Caucasians (15 percent) (Simopoulos 2004; Daar and Singer 2005).

It has also been suggested that susceptibility to obesity and type 2 diabetes may vary between populations on the basis of differences in genetic makeup (World Health Organization 2002). Chronic diseases are often preventable through lifestyle and dietary changes. Targeted dietary advice for subgroups as part of an overall approach to population-level disease prevention and health promotion would be one way to reduce health disparities (Eaton 2003). Such prevention strategies are cost-effective ways of avoiding the high social and economic costs of a treatment-based approach to chronic diseases (Doak 2002). According to the WHO, "small reductions in blood pressure, blood cholesterol and so on can dramatically reduce health costs" (Eaton 2003). To the extent that genetic contributions to Western lifestyle diseases are shared across industrialized and developing nations, there is promise of mutual benefit for both groups should they partner in research and development. Even in the absence of international research and development partnerships, the research done in industrialized countries can be shared relatively easily with developing countries, where there are similar patterns of genetic variability.

While noting that targeted dietary advice for susceptible populations is desirable, organizations such as WHO conclude that it is not feasible at present because many chronic diseases have multiple causes that are not yet well understood (World Health Organization 2003). They also note that given the rate of escalation of these diseases, changes in diet and lifestyle appears to be a much bigger causal factor than genetic susceptibility (Olden 2005).

There should be an effort to share nutrigenomic knowledge with developing countries, but to do so a number of issues will need to be addressed, including how best to finance and transfer technology (Fakuda-Parr et al. 2001; Singer and Daar 2001; Alwan and Modell 2003; Dowdeswell et al. 2003). In addition to knowledge transfer from industrialized to developing countries, nutrigenomic research could be done in developing countries in response to local health needs. For example, there might be an identifiable need for local research where it is suspected that genetic variations play a role in greater than average susceptibility to particular diseases in certain segments of the population. As noted above, it has long been suspected that increased susceptibility to obesity and diabetes among such groups as American Pima Indians, Australian Aborigines, and Pacific Islanders could be due to genetic causes (Caballero 2001).

6.4 NUTRIGENOMICS AND INTELLECTUAL PROPERTY

Issues of equity and access are affected by policies that determine the ownership, use, and distribution of gene sequences; information derived from gene sequences; genetic tests; and other supporting technologies, such as bioinformatics. In the early days of genome sequencing, considerable attention was paid to the potential of genomics for the improvement of health care. Some of this optimism, however, has been tempered by concerns that biotechnology-related intellectual property rights, as they are currently being applied, may in certain circumstances slow the realization of this potential (Fakuda-Parr et al. 2001). In particular, there are significant debates around the question of whether existing patent rights may limit public access to new tests, treatments, and research. Some evidence supports these concerns—at least in clinical testing (Merz et al. 2002)—suggesting that gene patents are hindering clinical research, interfering with patient care, and creating tensions among international trading partners (Alwan and Modell 2003). These concerns are being addressed domestically, for example, in Canada, and internationally by the World Health Organization (Canadian Biotechnology Advisory Committee 2006; World Health Organization 2006).

In recent years, advances in gene discovery have led to a large increase in the number of gene-related patents in both the commercial and public sectors. Two U.S. companies collectively filed more than 25,000 DNA-related patent applications by 2001 (Service 2001). This trend in DNA sequence patenting has given rise to a debate about how best to strike a balance between private ownership of genetic materials and technologies and public access for research and medical services. The patent system is seen as a way of attracting investment to the biotechnology sector, thereby fueling development of the next generation of biotechnology products. The competition protection created by the time-limited monopoly is meant to be balanced by the detailed disclosure of the invention to the public, thereby encouraging innovation (Arnold and Ogieslka-Zei 2002). This system is supposed to both provide investors with the chance to obtain returns from their investment, and lead to advances in science that will result in health benefits for the public. Some have pointed out that public investment in these innovations must also be factored in on grounds of fairness. For example, the public, through government funding, has contributed to the knowledge base resulting from the Human Genome Project, which is now being used to produce innovations in both the public and private sector. As contributors to the creation of the emerging benefits, the public has a claim to a share in those benefits. Others, however, point out that the bulk of the costs incurred in bringing a product based on basic research to market are incurred after the stage financed by public monies.

Concern is growing that the patent system may, in certain circumstances, act as an impediment to innovation, in particular in respect to clinical diagnostic tests, thus hindering improvements in health. Although patents are intended to create incentives for companies to develop products derived from the patented technologies and information, they restrict the availability of those technologies and information from broad use by others unless broadly licensed. Many of the patents currently being granted are licensed exclusively to single firms, allowing a strong monopoly in testing services related to the patented genes or genetic technologies. Patent monopolies and the costs associated with licensing threaten to reduce access to genetic testing and the resulting health benefits. According to Brad Sherman, a member of

---

Whose Genes Are They?

In the genetics field, the most widely known dispute over access to health care involves the claims of one company to hold patents to the BRCA1 and BRCA2 genes. Mutations of these genes significantly increase a woman's chance of developing inherited forms of breast and ovarian cancer. A U.S. company developed a genetic screening test for women and also claims to have discovered the BRCA2 gene. Where this mutation occurs, the company claims that it holds the patent on reading it.

Many countries have refused to honor the company's patent. In 2003, Ontario's health minister publicly announced that his government would ignore the company's patent and do its own tests for these mutations. According to the health minister, respecting the company's patents would greatly increase the cost of the tests, which in Ontario are paid for under a publicly funded health care system (*Cancer Weekly* 2003). France, Germany, the United Kingdom, and the Netherlands, whose laboratories have developed their own, cheaper tests for the breast cancer genes, have also protested the scope of the company's patent. These countries were being asked by the company to either send all test samples to its laboratory or to pay royalties for the right to use the genes (Wadman 2001). The European Parliament complained that honoring the U.S. biotechnology company's exclusionary patents would constitute an unwarranted restriction on their ability to deliver public health care (Fleck 2003). The European patents over BRCA1 and BRCA 2 were challenged in Europe. Although one of these challenges remains, one patent has been invalidated by the Patent Office and another has been greatly restricted in scope and thus effect.

---

the United Nations working group on intellectual property and biotechnology, evidence indicates that gene patents are having an adverse effect on the delivery of new genetic tests (GU News Service 2003).

There is growing concern that lack of access to DNA sequences and proprietary genetic databases could unduly delay or prevent much needed health research. Evidence is growing that commercial secrecy is hindering pure research by preventing the sharing of information (GU News Service 2003). When we look at the distribution of patents globally, this lack of access is even more troubling. Entities in the United States currently hold 80 percent of all gene sequence patents. The worry here is that future profits and resources will be concentrated in industrialized economies generally, and in the United States in particular, and that research and development will continue to be driven by the health care needs, demands, and markets of the industrialized world (World Health Organization 2002). This is mitigated, however, by the fact that patents over DNA sequences are rarely, if ever, sought and awarded in developing countries. This means that most DNA sequence patents have no effect on activities taking place solely within developing nations. It is only when an industrialized country entity provides services to a developing nation that problems can occur.

Intellectual property issues also arise in relation to genetic materials, information generated from those materials, technological processes and applications, and food products. There does appear to be a fair amount of patent activity in all these domains. One of the first companies to provide nutrigenomic tests, One Person Genetics, has filed a patent application for claims on intellectual property rights related to an automated screening and information management process dealing with relationships between genetic structural differences, the susceptibility of particular diseases, and metabolic pathways. It says that this also covers nutrition and medical advice that can reduce these risks and make therapy more effective. The company has also been granted a patent covering 2,500 genes related to metabolism and has filed patent applications "relating to the nutrition screening process and oligonucleotide sequences complementary to the polymorphisms of the 2,500 genes already patented and their use as probes in gene profiling technologies" (One Person Genetics 2003).

Whether these patents are valid remains to be seen. However, their sheer number gives one pause. It is likely that some of the 2,500 genes have been patented without their functions being known in the hope that a medical or scientific use will be discovered for them. This means that most will, if challenged, be held invalid. Nevertheless, given this "gene stockpiling" and fears that the patents will be held to be valid may prevent research by others. Patents related to screening and information management processes could also raise barriers to access for researchers, clinicians, and patients/consumers if not licensed. If the genes involved have relatively little potential for dramatic health effects, perhaps the public has less at stake, but the issue of limits imposed by intellectual property rights remains.

6.4.1   An Issue of Access to Scientific Information

Concerns about intellectual property barriers are increasing because of research interest in complex genetic diseases where multiple genes or gene–environment interactions are believed to be involved (Caulfield et al. 2006). Many serious illnesses, such as cardiovascular disease, some cancers, and diabetes, are now believed to be caused by many variant genes acting together. Consequently, a good deal of nutrigenomics research is shifting in this direction. This raises a problem that the patent system has yet to address, as existing rules have mostly been formulated in the context of single-gene disorders. The multiple genes involved in researching susceptibility to complex diseases might be patented by different firms, each of which would charge a licensing fee to gain access to the needed genes. The multiple licensing fees could make research prohibitively expensive for researchers by driving up the costs even before the product has been developed, ultimately hindering the kind of innovation needed for accessible health care. Whether patent pools, licensing societies, or other mechanisms such as those related to open sources can address the transaction costs involved with these many patents remains to be seen.

Proposed policy responses to the challenges posed by current patenting practices include banning gene patents—which would require an amendment to international trade agreements—making it more difficult to gain patent protection, narrowing the scope of patent protection, exempting noncommercial researchers from the reach of patents, encouraging patent pools, legislating compulsory licensing of important biotechnologies, recognizing the rights of third parties to oppose the granting of particular patents, and allowing the people who are the sources of the patented genes to have a greater say in their use (Chicago-Kent College of Law 2003). New models have also been put forward for intellectual property regimes which seek to strike a fairer balance between private and public interests and to incorporate social and ethical considerations into the intellectual property system (Gold et al. 2002).

Some have argued that gene patents ought not to be granted at all because patent rights to something as central to life as genes are inappropriate. They say that the human body should not be commodified, and DNA is part of the common heritage of humankind. Also, as noted above, there has been a significant public investment in these innovations: for example, through funding of the Human Genome Project. The knowledge from this research is now being used to produce innovations in both the public and private sectors. As contributors to the creation of the emerging benefits, the public can be seen to have claim to a share in those benefits, and models have been put forward to address the contributions made by those in developing countries who have contributed genetic resources to research and product development (Castle and Gold 2006).

6.5 CONCLUSION

Even though health disparities within and between countries continue to grow, the state of the science suggests that it is too early to predict whether the science of nutrigenomics and its applications will increase or decrease domestic or global health inequalities. In the meantime, knowledge of gene–nutrient interactions, family history, and molecular nutrition should help guide dietary advice for the prevention and management of disease. If implemented appropriately, nutrigenomics technologies can aid in reducing health inequalities and to ensure that disadvantaged people from all countries have reasonable access to its health benefits. In an effort to foster the potential benefits of nutrigenomics research and its applications, the following recommendations might be considered.

First, nutrigenomics should be subject to periodic review for their potential to help reduce health inequalities and to ensure that disadvantaged people globally have reasonable access to health benefits from emerging discoveries and applications. Nutrigenomic research should be responsive to the local health needs and priorities of developing countries. This can be achieved by ensuring, within a reasonable time, that there are effective, widely available, and affordable interventions at the disposal of the population being tested. Recommended health interventions should maximize local food availability and be consonant with local food culture.

Second, where nutrigenomics research leads to information that could be used to improve health in developing countries, we encourage the establishment of governance systems to ensure the appropriate transfer and implementation of these tools and knowledge to developing countries. Where evidence suggests that nutrigenomics testing and research will lead to improvements in health, intellectual property protection policies should aim to maximize the achievement of this goal while giving due consideration to other legitimate intellectual property considerations. The intellectual property system should also be designed to ensure that nutrigenomics discoveries, which are proven to be health related, are accessible to all who need them to the greatest extent reasonably possible.

REFERENCES

Acharya, T., A. S. Daar, H. Thorsteinsdottir, E. Dowdeswell, and P. Singer. 2004. Strengthening the role of genomics in global health. *PLos Medicine* 1:195–197.

Alwan, A., and B. Modell. 2002. Recommendations for introducing genetics services in developing countries. *Nature Review: Genetics* 4(1):61–68.

Annan, K. 2003. A challenge to the world's scientists. *Science.* 299:1485.

Arnold, Beth E., and Eva Ogielska-Zei. 2002. Patenting genes and genetic research tools: good or bad for innovation? *Annual Review in Genomics and Human Genetics* 3:415–432.

Beaglehole, R., and D. Yach. 2003. Globalization and the prevention and control of non-communicable diseases: the neglected chronic diseases of adults. *Lancet* 362:903–908.

Benatar, S. R., A. S. Daar, and P. A. Singer. 2003. Global health ethics: the rationale for mutual caring. *International Affairs* 1:107–138.

Bhardwaj, M., and D. R. J. Macer. 2003. Policy and ethical issues in applying medical biotechnology in developing dountries. *Medical Science Monitor* 9:RA52.

Braveman, P., B. Starfield, H. J. Geiger, and C. J. L. Murray. 2001. World health report 2000: How it removes equity from the agenda for public health monitoring and policy commentary: comprehensive approaches are needed for full understanding. *British Medical Journal* 323:678–681.

Caballero, B. 2001. Introduction: symposium on obesity in developing countries—biological and ecological factors. *Journal of Nutrition* 131:866S–870S.

Canadian Biotechnology Advisory Committee. 2006. *A Report by CBAC's Expert Working Party on Human Genetic Materials, Intellectual Property and the Health Sector.* Ottawa: CBAC.

Cancer Weekly. 2003. Biotech firm in dispute with Canada over genetic tests (Myriad Genetics). *Cancer Weekly*, January 28:25.

Castle, D., and E. R. Gold. 2006. Traditional knowledge and benefit sharing: from compensation to transaction. In *Accessing and Sharing the Benefits of the Genomics Revolution*, edited by P. Phillips and C. Onwuekwe. Dordrecht, The Netherlands: Springer-Verlag.

Caulfield, T., L. Sheremeta, E. R. Gold, J. F. Merz, and D. Castle. 2006. Informing genomic patent policy. In *Genetic Testing: Care, Consent and Liability*, edited by N. F. Sharpe and R. F. Carter. Hoboken, NJ: Wiley.

Chicago-Kent College of Law. 2003. Complex genetic disorders and intellectual property rights. Institute for Science, Law and Technology. Retrieved July 7, 2003, from `www.kentlaw.edu/islt/complexgene.html`.

Chopra, M., S. Galbraith, and I. Darnton-Hill. 2002. A global response to a global problem: the epidemic of overnutrition. *Bulletin of the World Health Organization* 80(12):952–958.

Daar, A. S., H. Thorsteinsdottir, D. K. Martin, A. C. Smith, S. Nast, and P. A. Singer. 2002. Top ten biotechnologies for improving health in developing countries. *Nature: Genetics* 32:229–232.

Darnton-Hill, I., B. Margets, and R. Deckelbaum. 2004. Public health nutrition and genetics: Implications for nutrition policy and promotion. *Proceedings of the Nutrition Society* 63:173–185.

Doak, C. 2002. Large-scale interventions and programmes addressing nutrition-related chronic diseases and obesity: examples from 14 countries. *Public Health Nutrition* 5:275–357.

Dowdeswell, E., A. S. Daar, and P. A. Singer. 2003. Bridging the genomics divide. *Global Governance* 9:1–6.

Eaton, L. 2003. Diet, nutrition and the prevention of chronic diseases. Geneva: World Health Organization.

Fine, M. J., S. A. Ibrahim, and S. B. Thomas. 2005. The role of race and genetics in health disparities research. *American Journal of Public Health* 95:2125–2128.

Fleck, F. 2003. No deal in sight on cheap drugs for poor countries. *Bulletin of the World Health Organization* 81:307–308.

Freimuth, V. S., S. C. Quinn, S. B. Thomas, G. Cole, E. Zook, and T. Duncan. 2001. African Americans' vies on research and the Tuskegee Syphilis Study. *Social Science and Medicine* 52:797–808.

Fukuda-Parr, S., S. Jahan, and H. Fu. 2001. *Human Development Report: Making New Technologies Work for Human Development.* Oxford: United Nations Development Programme.

Gold, E. R., D. Castle, L. M. Cloutier, A. S. Daar, and P. J. Smith. 2002. Needed: models of biotechnology intellectual property. *Trends in Biotechnology* 20(8):327–329.

GU News Service. 2003. Biotech advances create intellectual property minefield. Retrieved July 7, 2003, from www.gu.edu.au/text/er/news/2003_1/03feb28.

Gwatkin, D. R. 2000. Health inequalities and the health of the poor: What do we know? What can we do? *Bulletin of the World Health Organization* 78:3–18.

Harris, E., and M. Tanner. 2000. Health technology transfer. *British Medical Journal* 321:817–820.

Interleukin Genetics. 2006. Gensona™ genetic tests. Retrieved February 20, 2006, from www.ilgenetics.com/content/products-services/gensona.jsp.

Luis, J. 2002. A patent to kill? Comments on Resnik. *Developing World Bioethics* 2:86.

March of Dimes. 2006. *The March of Dimes Global Report on Birth Defects: The Hidden Toll of Dying and Disabled Children.* Atlanta, GA: MOD.

Merz, J. F., D. Magnus, M. K. Cho, and A. L. Caplan. 2002. Protecting subjects' interests in genetics research. *American Journal of Human Genetics* 70:965–971.

Morgan, S., J. Hurley, F. Miller, and M. Giacomini. 2003. Predictive genetic tests and health system costs. *Canadian Medical Association Journal* 168:989–993.

Olden, K. 2005. Health-related disparities: influence of environmental factors. *Medical Clinics of North America* 89:721–738.

One Person Genetics. 2003. Retrieved July 1, 2003, from www.onepersongenetics.com.

Pang, T. 2003. Equal partnership to ensure that developing countries benefit from genomics. *Nature Genetics* 33:18.

Powles, J., and N. Day. 2002. Interpreting the global burden of disease. *Lancet* 360:1342–1343.

Reddy, K. S. 2002. Cardiovascular diseases in the developing countries: dimensions, determinants, dynamics and directions for public health action. *Public Health Nutrition* 5:231–237.

Service, R. F. 2001. Proteomics: gene and protein patents get ready to go head to head. *Science* 294:2082–2083.

Simopoulos, A. P. 2004. Preface. In *Nutrigenetics and Nutrigenomics*, edited by A. P. Simopoulos, and J. M. Ordovas. Basel: Karger. VII–XII.

Singer, P. A., and A. S. Daar. 2001. Harnessing genomics and biotechnology to improve global health equity. *Science* 294:87–89.

Soo-Jin, S. 2005. Radicalizing drug design: implications of pharmacogenomics for health disparities. *American Journal of Public Health* 2005:2133–2138.

Trivedi, A. M., A. M. Zaslavsky, E. C. Schneider, and J. Z. Ayanian. 2005. Trends in the quality of care and racial disparities in Medicare managed care. *New England Journal of Medicine* 353:692–700.

University of California–Davis. 2003. New center will probe links between diet, genes and disease. NCMHD Center of Excellence for Nutritional Genomics. Retrieved 2003, from http://nutrigenomics.ucdavis.edu/pressarticles.htm.

Wadman, M. 2001. Testing time for gene patent as Europe rebels. *Nature* 413:443.

Wendler, D., R. Kington, J. Madans, G. Van Wye, H. Christ-Schmidt, L. A. Pratt, O. W. Brawley, C. P. Gross, and E. Emanuel. 2006. Are racial and ethnic minorities less willing to participate in health research? *Public Library of Science Medicine* 3:1–10.

Whitehead, M. 1992. The concepts and principles of equity and health. *International Journal of Health Services* 22:429–445.

World Bank. 2001. *World Development Report 2000/2001: Attacking Poverty.* London: WB.

World Health Organization. 2002. *Genomics and World Health: Report of the Advisory Committee on Health Research.* Geneva, Switzerland: WHO.

———. 2003. *Diet, Nutrition and the Prevention of Chronic Diseases.* Geneva, Switzerland: WHO.

———. 2006a. Commission on Intellectual Property Rights, Innovation and Public Health (CIPIH). Retrieved February 20, 2006, from www.who.int/intellectualproperty/en.

———. 2006b. *Preventing Chronic Diseases: A Vital Investment.* Geneva, Switzerland: WHO.

Yach, D., C. Hawkes, L. C. Gould, and K. J. Hofman. 2004. The global burden of chronic diseases: overcoming impediments to prevention and control. *Journal of the American Medical Association* 291:2616–2622.

Zimmet, P., K. G. Alberti, and J. Shaw. 2001. Global and societal implications of the diabetes epidemic. *Nature* 414:782–787.

7

CONCLUSIONS AND RECOMMENDATIONS

7.1 INTRODUCTION

I think there is a role for nutrigenomics in the future of health care if its benefits can be clearly shown and fairly applied.
—A response to the *Nutrition and Genes* public consultation paper

Nutrigenomics has great potential to provide significant, long-term health benefits through disease prevention and health promotion. As the science and its applications are developed, it is reasonable to anticipate continued private- and public-sector investments in nutrigenomic research and product and service development. Corresponding uptake by a public interested in personalized nutrition clearly linked to better health outcomes is also expected. As promising as these developments may seem at first glance, nutrigenomics faces significant challenges in terms of the development of the science and its applications. These are matched by challenges of a different type, the social determinants of nutrigenomics' success. Despite our belief about nutrigenomics' promise for the future, we have argued throughout this book that unless there is social acceptance of nutrigenomics, the field may stall.

We are not advocating for actions that will condition the social context so that nutritional genomics will succeed in the sense that we have an interest in nutrigenomics' success that would make us want to advocate on its behalf. Rather, we are arguing that if nutrigenomics is to succeed, a number of social considerations must be addressed. To our knowledge, this book is the first in-depth treatment of these social challenges for nutrigenomics. As mentioned in Chapter 1, our four-year-long study has involved working with an expert panel, focus groups, public feedback, and our own research and analysis. The conclusions and recommendations we have offered are meant for scientists, private interests, and regulators to integrate into their activities as they conduct research, create products and services, and regulate. In this chapter, we summarize those recommendations.

7.1.1   Nutrigenomic Science

> *I would want to know more about its efficacy and relevancy for specific disorders.*
>
> *I would want to know the detailed cause and effect that the food has on the gene. I would also want to fully understand the benefits or risks involved in changing my diet.*
> —Responses to the *Nutrition and Genes* public consultation paper

In Chapter 2 we argued that because the science of nutrigenomics and its nutrigenetic applications are rapidly evolving, there is a need to balance the speed of the field's evolution with a careful analysis of the quality of the science. We concluded that there must be an analysis of the risks and benefits of nutrigenetic interventions, based on systematic research in the form of well-designed controlled trials, cohort studies, or case–control studies. Because there is a paucity of high-quality nutritional intervention studies and the lack of information regarding the analytic and clinical validity of nutrigenetic tests, it is not currently possible to proclaim the overall utility of such tests. As a result, we recommend that:

1. Formal science and technology assessments should be conducted, and that these should be "rolling"—that is, conducted periodically to keep abreast of developments in the science and technology.
2. The results of this periodic science and technology assessment should be folded into the development of appropriate regulations and ethical guidelines which will ensure that the maximum benefits of nutritional genomics are made public and timely, with assurances that foreseeable risk has been mitigated and the public is protected from false claims.

7.1.2   Nutrigenomics and Health Information Management

As the insurance companies realize the validity and importance of some of the tests in preventing disease, major population groups who carry insurance will benefit.

Insurance companies must not be given access or be allowed to use this in any way. Everyone would end up with a high risk premium at some point regardless of their actual health.

—Responses to the *Nutrition and Genes* public consultation paper

Nutrigenomics requires the collection of biological samples, from which genetic information is derived. In Chapter 3 we discussed the potential to link genetic information and samples to personal information, including other medical information. This is a persistent ethical issue in research and clinical contexts involving genetic testing. Currently, nutrigenomic tests are viewed as a form of discretionary testing of a relatively nonsensitive nature, having mild to moderate informational impact. Breaches in the privacy of this information are therefore not thought to lead to immediate and significant harm, but this is not to say that people undergoing or conducting tests should have a cavalier attitude. The future potential impact of information mismanagement might be greater. Strategically speaking, it would be best if the highest standards for information management were adopted in nutrigenomics from the outset, rather than having to recover lost ground in the future. Therefore, we recommend that:

3. Those who offer nutrigenomic tests should follow applicable regulatory procedures and professional standards for obtaining informed consent, providing access to appropriate counseling, and protecting confidentiality and privacy.
4. Information from nutrigenomic tests should not be used in any way that causes discrimination against individuals, families, children, or vulnerable groups in either clinical or nonclinical contexts.
5. Nutrigenomic information should not be used by third parties, such as employers and insurers, unfairly to discriminate against people.

7.1.3   Nutrigenomic Service Delivery

I am concerned about over the counter genetic testing, misinformation and companies using the testing to promote and profit from the sale of supplements.

If nutrigenomics turns out to be as valuable for population-based health as predicted, anything short of offering it on a public health model would lead

to widespread inequities since only the wealthy would be able to afford testing and the appropriate diet.
                    —Responses to the *Nutrition and Genes* public consultation paper

If we assume that nutrigenomics is founded on good science and can deliver meaningful benefits to people, it seems incumbent to find means of providing access to nutrigenomic services. In Chapter 4 we do not settle the issue of who, among government, primary health care practitioners, or private companies, ought to be providing nutrigenomic services. Instead, we adopt the view that as the predictive value of nutrigenomic tests increases, there is a good argument to be made in favor of state support for public or private provision of nutrigenomics. How this will unfold in any particular jurisdiction is heavily dependent on a number of factors, such as prevailing health law, funding of health services, and the extent of the private health care market. Consequently, we have provided a generic rather than a jurisdiction-specific account of various service models and their strengths and weaknesses, and recommend:

6. The use of our four-model framework for analyzing potential service delivery models in a jurisdiction: the consumer model, the health practitioner model, a blended model, and the public health model.

7. When determining which approach will provide the best combination of service delivery and accountability, a number of factors should be considered, including the current state of the science, the predictive value of the tests, the number of trained personnel available, the levels of training and funding, and the degree of oversight and accountability.

8. There must be an analysis of the training required of health care professionals in the areas of genetics, nutritional science, and continual updating of the skills needed to link diet and health.

7.1.4 Regulation of Nutrigenomics

Uniform standards and policy are absolutely necessary. There is already too much misinformation and misconception surrounding the broad field of genetics. It would be dangerous to allow consumers, the market, stakeholders, etc. to regulate themselves; there is too much potential for misunderstanding and abuse.

I think that the border between nutrition and medicine is vanishing.

I think it would be too lax to regulate under foods, since nutrienomics has clear disease/health implications and involves testing. On the other hand, prescribing the consumption of a particular food (assuming it is a natural substance) is not the same as prescribing a drug. Thus I think that it would make more sense to create a new category of regulation that combines elements from both domains.

—Responses to the Nutrition and Genes public consultation paper

In Chapter 5 we considered the problem of developing appropriate regulations for a quickly evolving field such as nutrigenomics. Having noted that nutrigenomics-specific regulations do not exist, we pointed out that for some issues, such as the regulation of genetic tests, service delivery, and genetic antidiscrimination, some existing regulations might cover nutrigenomics. Other aspects of nutrigenomics, such as the potential for crossover between nutrigenomics and emerging regulations for supplements, functional foods, and nutraceuticals is more problematic. We argued that the way in which supplements, functional foods, and nutraceuticals are regulated will provide early models of regulation from which insights can be drawn for the future regulation of nutrigenomics. In addition, we considered how pharmaceutical-like bioactivity of some foods and food products would place nutrigenomics at the intersection of food and drug regulation. From this discussion we concluded that:

9. Regulators need to monitor developments in nutrigenomics in order to judge whether genetic tests to the public, delivery of nutrigenomic services, and potential misuses of genetic samples and profiles need to be regulated.

10. Regulators should be cognizant of changes in the regulation of the food industry that will affect nutrigenomics, especially the regulation of bioactive foods and food products such as supplements, functional foods, and nutraceuticals.

11. The need for new regulations for nutrigenomics must be monitored, particularly if nutrigenomics awkwardly straddles food and drug regulations. New approaches may need to be developed to cope with the regulatory convergence caused by nutrigenomics.

7.1.5 Access and Equity

Maybe it is not necessary that everybody has access. It may be considered as a luxury product.

Nutrigenomics is sophisticated and expensive technology—just the sort of thing that the "have-nots" will surely not have.

The issue is not unique to nutrigenomics. Social inequity and poverty are global problems.

— Responses to the Nutrition and Genes public consultation paper

One of the selling points of the Human Genome Project was that it would enable us to understand the common heritage of humankind and that applications flowing from it ought to benefit humankind in common. Yet the situation is dire for roughly 90 percent of the world's population. Nutrigenomics stands at a fork in the road: It can either be pressed into the service of the 600 million people living in affluent countries, or it can serve them and the rest of the world at the same time. If appropriately implemented, nutrigenomics technologies can aid in reducing health inequalities and to ensure that disadvantaged people from all countries have reasonable access to its health benefits. In an effort to foster the potential benefits of nutrigenomics research and its applications, we recommend that:

12. The field of nutrigenomics should be reviewed periodically to monitor how it can be more responsive to the local health needs and priorities of developing countries. There should be effective, widely available, and affordable interventions at the disposal of the population being tested.
13. Recommended health interventions should maximize local food availability and be consonant with local food culture.
14. Where nutrigenomics research leads to information that could be used to improve health in developing countries, governance systems to ensure the appropriate transfer and implementation of these tools and knowledge to developing countries should be created. In particular, intellectual property protection policies should aim to maximize the achievement of this goal.

7.2 A FINAL WORD

We conclude by noting that in addition to the opportunities and challenges discussed above, nutrigenomics poses other interesting challenges that we are unable to consider in detail in this book. For example, given the widespread interest in the applications of the Human Genome Project, there is a risk that we could put too much emphasis on our genes as the basis for our health status, and pay too little attention to social and environmental factors and their interplay with genes. Is this a renewed form of privileging genes, even though nutrigenomics is about understanding genes in light of their interaction with the ongoing exposure to nutrients? Equally, we might consider whether there is something disturbing about the continued "medicalization" of food, and whether cultural and aesthetic attributes of food are being supplanted with medical associations. Finally, we also raise the political issue of whether nutrigenomics is a science that is riding the wave of "personal responsibility" that has been sweeping through some industrialized countries as they have been led by neoconservative politics. In this respect it would be interesting to see if the role for nutrigenomics in socialized public health systems is smaller than right-of-center systems.

These are topics for further reflection. In this book, we have started the conversation about the significant scientific hurdles, the opportunities, and the challenges ahead of nutrigenomics. No doubt this first in-depth analysis of the social aspects of nutrigenomics will have its errors and omissions. We take these in stride, however, since a proactive bioethics approach mandates that such a "first look" be conducted now. We have tried to categorize and analyze the social issues the field will face and to provide recommendations that will guide the science, its applications, and its regulation. If such clarification can facilitate the ability of the field to deliver on its legitimate benefits, we believe that our investment of time in conducting this study, and yours in reading it, will have been worthwhile.

INDEX

161